Heat and Mass Transfer

Dr.R. Sivasubramanian, B.E., M.E., Ph.D.,

Principal,

Latha Mathavan Engineering College,

Kidaripatti, Madurai.

Published by

Second Edition: March 2019

ISBN 978-93-86638-71-7

Author

Dr.R. Sivasubramanian

Bonfring

309, 2nd Floor, 5th Street Extension, Gandhipuram,

Coimbatore-641 012.

Tamilnadu, India.

E-mail: info@bonfring.org

Website: www.bonfring.org

Phone: 0422 4213231

Preface

*I am glad to present the book entitled, "**Heat and Mass Transfer**" to the engineering students of Mechanical. The course contents have been planned in such a way that the general requirements of all engineering students are fulfilled.*

During my long experience of teaching this subject to undergraduate engineering students for the past 15 years. I have observed that the students face difficulty in understanding clearly the basic principles, fundamental concepts and theory without adequate solved problems along with the text. To meet this very basic requirement to the students, a large number of the questions taken from the examinations have been solved along with the text in M.K.S. and S.I. Units.

The book is written in a simple and easy-to-follow language, so that even an average student can grasp the subject by self-study.

I am thankful to my colleagues, friends and students who encouraged me to write this book. I express my appreciation and gratefulness to my publisher for his most cooperative, painstaking attitude and untiring efforts for bringing out the book in a short period.

Though every care has been taken in checking the manuscripts and proof reading, yet claiming perfection is very difficult. I shall be very grateful to the readers and users of this book for pointing any mistakes that might have crept in. Suggestions for improvement are most welcome and would be incorporated in the next edition with a view to make the book more useful.

Dr.R. Sivasubramanian

UNIT 1

CONDUCTION

1.1. Heat Transfer

Heat transfer can be defined as the transmission of energy from one region to another region due to temperature difference.

1.1.1. Modes of Heat Transfer

- Conduction
- Convection
- Radiation

Tabulation 1.1.1

S.No	Description	Conduction	Convection	Radiation
1	Definition	Heat conduction is a mechanism of heat transfer from a high temperature region to low temperature region within a medium (Solid, liquid, gases) or between medium in direct physical contact.	Convection is a process of heat transfer that will occur between a solid surface and a fluid medium when they are at different temperatures.	The heat transfer from one body to another without any transmitting medium.
	Medium	Pure conduction is found only insolids.	It is possible only in the presence of fluid medium.	It is an electromagnetic wave phenomenon.
2	Name of Law	Fourier law	Newton's law	Stefan-Boltzman law
3	Law	The rate of heat transfer is directly proportional to particular area, at a particular distance, transfer of heat from high to lower.	Heat transfer from the moving fluid to solid surface is given by the equation.	The emissive power of a black body is proportional to the fourth power of absolute temperature.
4	Formulas	$Q \alpha - A \dfrac{dT}{dx}$	$Q \alpha\, A\, \Delta T$	$E_b \alpha T^4$, $Q_{rad} = \sigma A T^4$
		$Q = - kA \dfrac{dT}{dx}$	$Q = hA\,(T_w - T_\infty)$	$E_b \alpha T^4$, $Q_{rad} = \varepsilon \sigma A(T_1^4 - T_2^4)$
5	Notation	K – Thermal conductivity in w/mk	h – Local heat transfer co-efficient in $w/m^2 k$	σ = Stefan – boltzman constant in 5.67 x 10^{-8} $w/m^2 k^4$ ε = Emissivity

1.1.2. Fourier Law of Conduction

The rate of heat transfer is directly proportional to in particular area, at a particular distance transfer of heat from high to lower.

$$Q \ \alpha \quad - A \ \frac{dT}{dx}$$

$$Q \ = \quad - KA \ \frac{dT}{dx}$$

[(-) Negative sign indicates the temperature decrease high to lower]

where,

Q	-	Heat flow in W
K	-	Thermal conductivity in w/mk
$\dfrac{dT}{dx}$	-	Temperature gradient
A	-	Area in m².

1.1.3. General Heat Conduction Equation in Cartesian Coordinates

Step 1

Consider a small rectangular element of side dx, dy, dz as shown in Fig.1.1.

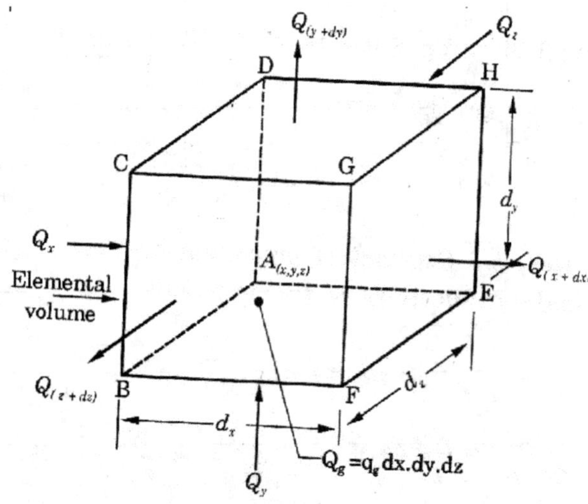

Figure 1.1: Cartesian Coordinates

Step 2

The energy balance equation as follows,

$$\begin{bmatrix} Net\ heat\ conducted\ int\,o \\ element \\ dx\ dy\ dz\ per\ unit\ time \end{bmatrix}(I)\ +\ \begin{bmatrix} Internal\ heat\ generated \\ per\ unit\ time \end{bmatrix}(II)\ =$$

$$\begin{bmatrix} Increase\ in\ int\,ernal\ energy \\ per\ Unit\ energy\ per\ unit\ time \end{bmatrix}(III)\ +\ \begin{bmatrix} Work\ done\ by\ element \\ per\ unit\ time \end{bmatrix}(IV) \rightarrow (1)$$

Step 3

(i) Net heat conducted in to element dx dydz per unit time

Fourier law of heat conduction

$$Q_x = -k_x\,A\,\frac{\partial T}{\partial x}$$

$$Q_x = -k_x . \frac{\partial T}{\partial x} . dy.\ dz. \qquad\qquad \rightarrow \qquad (1.1)$$

The rate of heat flow out of the element in x direction.

$$Q_{x+dx} \qquad = Q_x + \frac{\partial Q_X}{\partial x}\,dx$$

$$= -k_x. \frac{\partial T}{\partial x}\,dy.dz + \frac{\partial}{\partial x}\left[-k_x. \frac{\partial T}{\partial x}\,dx.dy.dz\right]$$

$$Q_{x+dx} \qquad = -k_x. \frac{\partial T}{\partial x}\,dy.dz - \frac{\partial}{\partial x}\left[k_x. \frac{\partial T}{\partial x}\,dx.dy.dz\right] \qquad \rightarrow \qquad (1.2)$$

Subtracting eqn (1.1) – equ (1.2)

$$Q_x - Q_{x+dx} = -k_x \frac{\partial T}{\partial x}\,dy.dz + k_x. \frac{\partial T}{\partial x}\,dy.dz + \frac{\partial}{\partial x}\left[k_x. \frac{\partial T}{\partial x}\,dx.dy.dz\right]$$

$$Q_x - Q_{x+dx} = \frac{\partial}{\partial x}\left[k_x. \frac{\partial T}{\partial x}\,dx.dy.dz\right] \qquad \rightarrow \qquad (1.3)$$

Similarly,

$$Q_y - Q_{y+dy} = \frac{\partial}{\partial y}\left[k_y. \frac{\partial T}{\partial y}\,dx.dy.dz\right] \qquad \rightarrow \qquad (1.4)$$

$$Q_z - Q_{z+dz} = \frac{\partial}{\partial z}\left[k_z. \frac{\partial T}{\partial z}\,dx.dy.dz\right] \qquad \rightarrow \qquad (1.5)$$

$\therefore$ Net heat conducted Adding (1.3) + (1.4) + (1.5)

$$\frac{\partial}{\partial x}\left[k_x.\frac{\partial T}{\partial x}\,dx\,dy\,dz\right] + \frac{\partial}{\partial y}\left[k_y.\frac{\partial T}{\partial y}\,dx\,dy\,dz\right] + \frac{\partial}{\partial z}\left[k_x.\frac{\partial T}{\partial z}\,dx\,dy\,dz\right]$$

Material can be consider as isotropic, so which, $k_x = k_y = k_z = k$

$$k\left[\frac{\partial^2 T}{\partial x^2}+\frac{\partial^2 T}{\partial y^2}+\frac{\partial^2 T}{\partial z^2}\right]dx\,dy\,dz \qquad\qquad \rightarrow \qquad (1.6)$$

Step 4

(ii) Internal heat generated per unit time

$$= q^{\bullet}\,dx\,dy\,dz \qquad\qquad\qquad\qquad \rightarrow \qquad (1.7)$$

Step 5

(iii) Increasing in internal energy per unit time

$$= m\,C_p\frac{\partial T}{\partial t}$$

$$= = \rho.c_p.\,dx\,dy\,dz.\frac{\partial T}{\partial t} \qquad\qquad [\therefore \text{mass} = \text{Density x volume}]$$

$$= \rho.\,c_p.\frac{\partial T}{\partial t}\,dx\,dy\,dz \qquad\qquad\qquad \rightarrow \qquad (1.8)$$

Substituting equation (1.6), (1.7) and (1.8) in eqn (1)

Step 6

$$k\left[\frac{\partial^2 T}{\partial x^2}+\frac{\partial^2 T}{\partial y^2}+\frac{\partial^2 T}{\partial z^2}\right]dx\,dy\,dz + q^{\bullet}\,dx\,dy\,dz = \rho.cp.\frac{\partial T}{\partial t}\,dx\,dy\,dz$$

$$k\left[\frac{\partial^2 T}{\partial x^2}+\frac{\partial^2 T}{\partial y^2}+\frac{\partial^2 T}{\partial z^2}\right]+q^{\bullet} = \rho.c_p.\frac{\partial T}{\partial t}$$

Both sides divided by k

$$\frac{\partial^2 T}{\partial x^2}+\frac{\partial^2 T}{\partial y^2}+\frac{\partial^2 T}{\partial z^2}+\frac{q^{\bullet}}{k}=\frac{\rho.c_p}{k}\frac{\partial T}{\partial t}$$

$$\frac{\partial^2 T}{\partial x^2}+\frac{\partial^2 T}{\partial y^2}+\frac{\partial^2 T}{\partial z^2}+\frac{q^{\bullet}}{k}=\frac{1}{\alpha}\frac{\partial T}{\partial t} \qquad\qquad \rightarrow \qquad (1.9)$$

where,

$$\alpha = \text{Thermal diffusivity} = \frac{k}{\rho.C_p} \text{ - m}^2/s$$

It is a general three dimensional heat conduction equation in Cartesian coordinates. The equation (1.9) is known as Fourier biot equation.

Case (i)

3D and 1D poissons equation

$$[\text{Steady state } \frac{\partial T}{\partial t} = 0]$$

The steady state heat conduction equation with heat generation is called poisson equation.

$$[\text{i.e. } \frac{\partial T}{\partial t} = 0]$$

Fourier – Biot equation reduces to

$$\frac{\partial^2 T}{\partial x^2} + \frac{\partial^2 T}{\partial y^2} + \frac{\partial^2 T}{\partial z^2} + \frac{q^\bullet}{k} = 0 \qquad \text{... 3D poisson equation}$$

$$\frac{\partial^2 T}{\partial x^2} + \frac{q^\bullet}{k} = 0 \qquad \text{... 1D poisson equation}$$

Case (ii)

3D and 1D Fourier's equation

[Transient state and no heat generation $q^\bullet = 0$]

$$\frac{\partial^2 T}{\partial x^2} + \frac{\partial^2 T}{\partial y^2} + \frac{\partial^2 T}{\partial z^2} = \frac{1}{\alpha} \cdot \frac{\partial T}{\partial t} \qquad \text{... 3D poisson equation}$$

$$\frac{\partial^2 T}{\partial x^2} = \frac{1}{\alpha} \frac{\partial T}{\partial t} \qquad \text{... 1D poisson equation}$$

Case (iii)

3D and 1D Laplace equation

[Steady state and No Heat Generation]

$$\frac{\partial T}{\partial t} = 0 \text{ and } q^\bullet = 0$$

$$\frac{\partial^2 T}{\partial x^2} + \frac{\partial^2 T}{\partial y^2} + \frac{\partial^2 T}{\partial z^2} = 0 \qquad\qquad \ldots \text{3D poisson equation}$$

$$\frac{\partial^2 T}{\partial x^2} = 0 \qquad\qquad \ldots \text{1D poisson equation}$$

1.1.4. General Heat Conduction Equation in Polar Co-ordinates (Cylindrical Co-ordinates)

Step 1

Consider a small cylindrical element of side dr, dϕ and dz as shown in fig.1.2

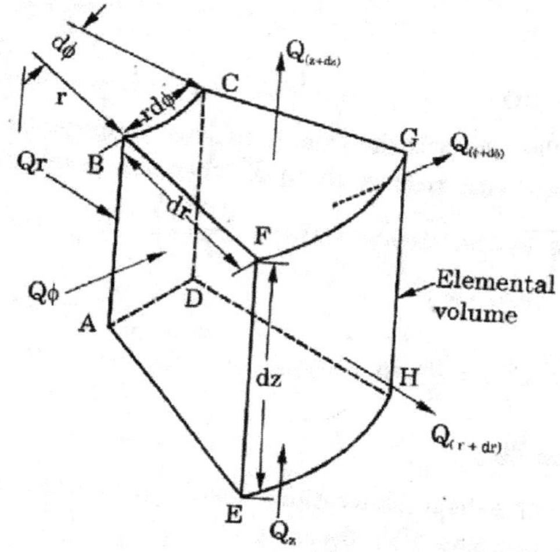

Figure 1.2: Cylindrical Co-ordinates

Volume of the element dv = r dϕ dr dz.

Step 2

The energy balance of cylindrical element is obtained from first law of thermodynamics.

$$
\begin{bmatrix}
\text{Net heat conducted} \\
\text{into element from} \\
\text{all the coordinates} \\
\text{directions}
\end{bmatrix}
+
\begin{matrix}
\text{Heat Generated} \\
\text{within the element}
\end{matrix}
=
\begin{matrix}
\text{Heat stored in} \\
\text{the element}
\end{matrix}
\qquad \rightarrow (1)
$$

Step 3

(i)   Net heat conducted into element from all the co-ordinates directions.

Plane (r,ϕ)	plane (ϕ,z)	plane (z,r)
Heat entering in the element (r,ϕ) plane in time dθ	Heat entering in the element (p, z) plane in time dθ	Heat entering in the element (z, r) plane in time dθ
$Q_z = -k(dr\ rd\phi)\left(\dfrac{\partial T}{\partial z}\right)d\theta$ Heat leaving from the element through (r,ϕ) plane in time dθ	$Q_r = -k\ (rd\phi\ dz)\dfrac{\partial T}{\partial r}\ d\theta$ Heat leaving from the element through (ϕ, z) plane in time dθ	$Q_\phi = -k(dr.dz)\left(\dfrac{\partial T}{r.\partial\phi}\right)d\theta$ Heat leaving from the element through (z, r) plane in time dθ
$Q_{z+dz} = Q_z + \dfrac{\partial}{\partial z}\ Q_z\ dz$ Net heat conducted into the element through (r,ϕ) plane in time dθ	$Q_{r+dr} = Q_r + \dfrac{\partial Q_r}{\partial r}\ dr$ Net heat conducted in to the element through (ϕ, z) plane in time dθ	$Q_{\phi+d\phi} = Q_\phi + \dfrac{\partial Q_\phi}{rd\phi}\ rd\phi$ Net heat conducted in to the element through (z, r) plane in time dθ
$= Q_z - Q_{z+dz}$	$= Q_r - Q_{r+dr}$	$= Q_\phi - Q_{\phi+d\phi}$
$= -\dfrac{\partial}{\partial z}\ (Q_z)\ dz$	$= -\dfrac{\partial}{\partial r}\ (Q_r)\ dr$	$= -\dfrac{\partial}{rd\phi}\ (Q_\phi)\ rd\phi$
$= -\dfrac{\partial}{\partial z}\left[-k(dr\ rd\phi)\left(\dfrac{\partial T}{\partial z}\right)d\theta\right]dz$	$= -\dfrac{\partial}{\partial r}\left[-k(rd\phi dz)\left(\dfrac{\partial T}{\partial r}\right)d\theta\right]dr$	$= -\dfrac{\partial}{rd\phi}\left[-k(drdz).\left(\dfrac{\partial T}{r.\partial\phi}\right)d\theta\right].rd\phi$
$= k\left(\dfrac{\partial^2 T}{\partial z^2}\right)(dr.rd\phi.dz).d\theta$ $\rightarrow$ (1.10)	$= k(drrd\phi dz)\dfrac{\partial}{\partial r}(r.\dfrac{\partial T}{\partial r})\ d\theta$ $= k(drrd\phi dz)\left(\dfrac{\partial^2 T}{\partial r^2} + \dfrac{1}{r}\dfrac{\partial T}{\partial r}\right)d\theta$ $\rightarrow$ (1.11)	$= k\dfrac{\partial}{rd\phi}\left[\dfrac{1}{r}\dfrac{\partial T}{\partial\phi}\right]d\theta(dr\ rd\phi dz)$ $= k\left[\dfrac{1}{r^2}\dfrac{\partial^2 T}{\partial\phi^2}\right](dr\ rd\phi dz)d\theta$ $\rightarrow$ (1.12)

Adding equation (1.10) + (1.11) + (1.12)

$$= k\frac{\partial^2 T}{\partial z^2}(dr\ rd\phi dz)\ d\theta + k(dr\ rd\phi dz)\left[\frac{\partial^2 T}{\partial r^2} + \frac{1}{r}\frac{\partial T}{\partial r}\right]d\theta + k\left[\frac{1}{r^2}\frac{\partial^2 T}{\partial\phi^2}\right](dr\ rd\phi dz)\ d\theta$$

$$= k\ (dr\ rd\phi\ dz)\ d\theta\left[\frac{\partial^2 T}{\partial Z^2} + \frac{\partial^2 T}{\partial r^2} + \frac{1}{r}\frac{\partial T}{\partial r} + \frac{1}{r^2}\frac{\partial^2 T}{\partial\phi^2}\right]$$

$$= k\ (dr\ rd\phi\ dz)\ d\theta\left[\frac{\partial^2 T}{\partial r^2} + \frac{1}{r}\frac{\partial T}{\partial r} + \frac{1}{r^2}\frac{\partial^2 T}{\partial\phi^2} + \frac{\partial^2 T}{\partial Z^2}\right] \qquad\rightarrow\qquad (1.13)$$

Step 3

(ii) Heat generated within the element

$$Q = q^{\bullet} \, (dr \, rd\phi \, dz) \, d\theta \qquad\qquad \rightarrow \qquad (1.14)$$

Step 4

(iii) Heat stored in the element

Increase in internal energy = Net heat stored in the element

$$= \rho \, (dr \, rd\phi \, dz) \, c_p . \frac{\partial T}{\partial \theta} \, .d\theta \qquad\qquad \rightarrow \qquad (1.15)$$

Step 5

Substituting equation (1.13) (1.14) and (1.15) in eqn (1)

$$k(dr \, rd\phi \, dz) \, d\theta \left[\frac{\partial^2 T}{\partial r^2} + \frac{1}{r} \frac{\partial T}{\partial r} + \frac{1}{r^2} \frac{\partial^2 T}{\partial \phi^2} + \frac{\partial^2 T}{\partial Z^2} \right] + q^{\bullet} \, (dr \, rd\phi \, dz) d\theta$$

$$= \rho \, (dr \, rd\phi \, dz) \, d\theta \, \frac{\partial T}{\partial \theta}$$

Divided by $(dr \, rd\phi \, dz) \, d\theta$

$$\Rightarrow k \left[\frac{\partial^2 T}{\partial r^2} + \frac{1}{r} \frac{\partial T}{\partial r} + \frac{1}{r^2} \frac{\partial^2 T}{\partial \phi^2} + \frac{\partial^2 T}{\partial Z^2} \right] + q^{\bullet} \quad = \quad \rho . \, c_p \, \frac{\partial T}{\partial \theta}$$

Divided by k

$$\Rightarrow \left[\frac{\partial^2 T}{\partial r^2} + \frac{1}{r} \frac{\partial T}{\partial r} + \frac{1}{r^2} \frac{\partial^2 T}{\partial \phi^2} + \frac{\partial^2 T}{\partial Z^2} \right] + \frac{q^{\bullet}}{k} \quad = \quad \frac{\rho . c_p}{k} . \frac{\partial T}{\partial \theta}$$

$$\Rightarrow \left[\frac{\partial^2 T}{\partial r^2} + \frac{1}{r} \frac{\partial T}{\partial r} + \frac{1}{r^2} \frac{\partial^2 T}{\partial \phi^2} + \frac{\partial^2 T}{\partial Z^2} \right] + \frac{q^{\bullet}}{k} \quad = \quad \frac{1}{\alpha} . \frac{\partial T}{\partial \theta} \qquad \rightarrow \qquad (1.16)$$

$(\therefore \alpha = k/\rho c_p)$

It is a general three dimension heat conduction equation in cylindrical co-ordinates.

Case (i)

3D and 1D poissons Equation (Steady state $\frac{\partial T}{\partial \theta} = 0$)

$$\left[\frac{\partial^2 T}{\partial r^2} + \frac{1}{r} \frac{\partial T}{\partial r} + \frac{1}{r^2} \frac{\partial^2 T}{\partial \phi^2} + \frac{\partial^2 T}{\partial Z^2} \right] + \frac{q^{\bullet}}{k} = 0 \qquad \text{.... 3D Poisson eqn}$$

$$\frac{\partial^2 T}{\partial r^2} + \frac{q^{\bullet}}{k} = 0 \qquad\qquad \text{.... 1D Poisson eqn}$$

Case (ii)

3D and 1D Fourier equation

(Transient state and No heat generation $q^{\bullet} = 0$]

$$\left[\frac{\partial^2 T}{\partial r^2} + \frac{1}{r}\frac{\partial T}{\partial r} + \frac{1}{r^2}\frac{\partial^2 T}{\partial \phi^2} + \frac{\partial^2 T}{\partial Z^2}\right] = \frac{1}{\alpha}\frac{\partial T}{\partial \theta} \qquad \ldots \text{3D Fourier's eqn}$$

$$\frac{\partial^2 T}{\partial r^2} = \frac{1}{\alpha}\frac{\partial T}{\partial t} \qquad \ldots \text{1D Fourier's eqn}$$

Case (iii)

3D and 1D Laplace equation

[steady state and No heat generation] $\left[\dfrac{\partial T}{\partial t} = 0 \text{ and } q^{\bullet} = 0\right]$

$$\left[\frac{\partial^2 T}{\partial r^2} + \frac{1}{r}\frac{\partial T}{\partial r} + \frac{1}{r^2}\frac{\partial^2 T}{\partial \phi^2} + \frac{\partial^2 T}{\partial Z^2}\right] = 0 \qquad \ldots \text{3D Laplace eqn}$$

$$\frac{\partial^2 T}{\partial r^2} = 0 \qquad \ldots \text{1D Laplace eqn}$$

1.1.5. One Dimensional Steady State Heat Conduction through Plane Wall

Step 1

Consider a plane wall of uniform thermal conductivity k, thickness L with inner temperature T_1 and outer temperature T_2. Let us consider a small element area of thickness 'dx'.

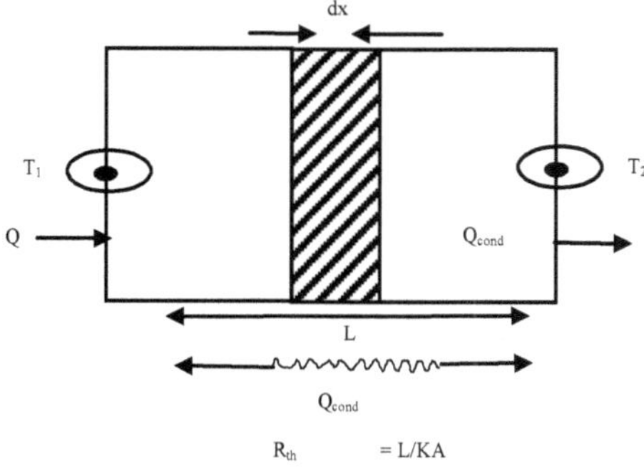

Figure 1.3: Heat Conduction through Plane Wall

Step 2

3D general heat conduction equation in Cartesian coordinates is given by

$$\frac{\partial^2 T}{\partial x^2} + \frac{\partial^2 T}{\partial y^2} + \frac{\partial^2 T}{\partial z^2} + \frac{\overset{\bullet}{q}}{k} = \frac{1}{\alpha}\frac{\partial T}{\partial t} \qquad \rightarrow (1)$$

Step 3

(i) Steady state condition $\dfrac{\partial T}{\partial t} = 0$ and

(ii) One dimensional heat flow in x direction

$$\frac{\partial^2 T}{\partial y^2} = 0, \ \frac{\partial^2 T}{\partial z^2} = 0$$

(iii) No heat generation

$$\frac{\overset{\bullet}{q}}{k} = 0$$

$\qquad \therefore \quad$ Equation (1)

$$\therefore \frac{\partial^2 T}{\partial x^2} = 0 \text{ (one dimensional eqn)} \qquad\qquad \rightarrow \qquad (1.17)$$

Step 4

Integrate on both sides in eqn. 1.17

$$\frac{\partial T}{\partial x} = C_1$$

$$\partial T = C_1 \partial x$$

$$dT = C_1\, dx \qquad\qquad\qquad \rightarrow \qquad (1.18)$$

Again integrate on both sides in eqn.1.18

$$T = C_1 x + C_2 \qquad\qquad\qquad \rightarrow \qquad (1.19)$$

Step 5

Apply boundary conditions

Boundary Condition (1)	Boundary Condition (2)
$T = T_1$	$T = T_2$
$x = 0$	$x = L$

Using (1) boundary condition in eqn (1.19)

$$T_1 = C_1\,(o) + C_2$$

$T_1 = C_2$

$C_2 = T_1$

Using (2) boundary condition in eqn (1.19)

$T_2 = C_1 (L) + C_2$

$T_2 = C_1 . L + T_1$

$$C_1 = \frac{T_2 - T_1}{L}$$

Step 6

Substituting value of C_1 and C_2 in eqn

$$T = \frac{T_2 - T_1}{L} x + T_1 \qquad\qquad \rightarrow \qquad (1.20)$$

Step 7

Fourier's law of heat conduction equation as follows

$$Q = -kA \frac{dT}{dx}$$

$Q.dx = -kA\, dT$

Both sides integrate

$$Q. \int_{o}^{L} dx = -kA \int_{T_1}^{T_2} dT$$

$$Q\left[x\right]_{0}^{L} = -kA\left[T\right]_{T_1}^{T_2}$$

$Q[L-o] = -kA\,(T_2 - T_1)$

$QL = kA\,(T_1 - T_2)$

$$Q = \frac{kA}{L} (T_1 - T_2)$$

$$Q = \frac{T_1 - T_2}{L / kA} \qquad\qquad \rightarrow \qquad (1.21)$$

$$Q = \frac{T_1 - T_2}{R}$$

$$Q = \frac{\Delta T_{overall}}{R} \quad \text{where, Thermal resistance of slab } [\therefore R = L/kA]$$

$\Delta T = T_1 - T_2$

1.1.6. *One Dimensional Heat Conduction through Hollow Cylinder*

Step 1

Consider a hollow cylinder of inner radius r_1, outer radius r_2, inner temperature T_1, outer Temperature T_2 and thermal conductivity k. Let us consider a small elemental area of thickness 'dr'.

Step 2

3D heat conduction equation through cylinder is given by,

$$\frac{d}{dr}\left(r\frac{dT}{dr}\right) = 0 \qquad\qquad \rightarrow \qquad (1.22)$$

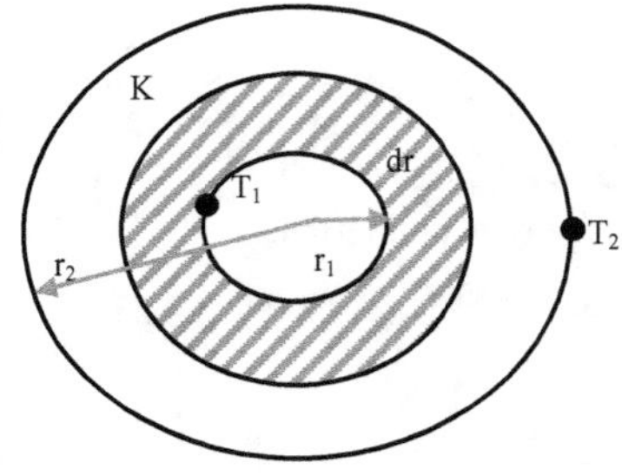

Figure 1.4: Heat Conduction through Hollow Cylinder

Step 3

Integrate on Both sides in eqn (1.22)

$$\int\frac{d}{dr}(r.\frac{dT}{dr}) = \int 0$$

$$r.\frac{dT}{dr} = C_1$$

$$\frac{dT}{dr} = \frac{C_1}{r}$$

Again integrate above equation

$$dT = C_1 . \frac{dr}{r}$$

$$\int dT = C_1 \int \frac{dr}{r}$$

$$T = C_1 \, ln(r) + C_2 \qquad\qquad \rightarrow \qquad (1.23)$$

Step 4

Boundary Condition (1)	Boundary Condition (2)
$T = T_1$	$T = T_2$
$r = r_1$	$r = r_2$

Applying boundary condition (1) and (2) in eqn (1.23)

$$T_1 = C_1 ln\,(r_1) + C_2 \qquad\qquad \rightarrow \qquad (1.24)$$

$$T_2 = C_1 ln\,(r_2) + C_2 \qquad\qquad \rightarrow \qquad (1.25)$$

Equation (1.24) – (1.25)

$$T_1 - T_2 = (C_1 ln\,(r_1) + C_2) - (C_1 ln\,(r_2) + C_2)$$

$$T_1 - T_2 = C_1 ln\,(r_1) + C_2 - C_1 ln\,(r_2) - C_2$$

$$T_1 - T_2 = C_1\,[ln\,(r_1) - ln\,(r_2)]$$

$$T_1 - T_2 = C_1 ln\left(\frac{r_1}{r_2}\right)$$

$$C_1 = \frac{T_1 - T_2}{ln\left(\dfrac{r_1}{r_2}\right)}$$

C_1 value substituting in eqn (1.24)

$$T_1 = C_1 ln\,(r_1) + C_2$$

$$T_1 = \frac{T_1 - T_2}{ln\left(\dfrac{r_1}{r_2}\right)} ln\,(r_1) + C_2$$

$$C_2 = \left[T_1 - \frac{T_1 - T_2}{ln\left(\dfrac{r_1}{r_2}\right)} ln\,(r_1) \right]$$

Step 5

Substituting value of C_1 and C_2 in eqn (1.23)

$$T = C_1 ln\,(r) + C_2$$

$$T = \frac{T_1 - T_2}{ln\left(\dfrac{r_1}{r_2}\right)} ln\,(r) + \left[T_1 - \frac{T_1 - T_2}{ln\left(\dfrac{r_1}{r_2}\right)} ln\,(r_1) \right] \qquad \rightarrow \qquad (1.26)$$

Step 6

From Fourier law of conduction

$$Q = -kA . \frac{dT}{dr}$$

$$Q.dr = -kA \, dT$$

$$Q.dr = -k \, (2\pi rL) \, dT$$

$$Q . \frac{dr}{r} = -2\pi kL . dT$$

Both sides integrate

$$Q \int_{r_1}^{r_2} \frac{dr}{r} = -2\pi kL \int_{T_1}^{T_2} dT$$

$$Q \ln (r)_{r_1}^{r_2} = -2\pi kL \, [T]_{T_1}^{T_2}$$

$$Q \ln (r_2 - r_1) = -2\pi kL(T_2 - T_1)$$

$$Q \ln \left(\frac{r_2}{r_1}\right) = 2\pi kL(T_1 - T_2)$$

$$Q = \frac{2\pi kL}{\ln \left(\dfrac{r_2}{r_1}\right)} (T_1 - T_2)$$

$$Q = \frac{T_1 - T_2}{\dfrac{\ln \left(r_2 / r_1\right)}{2\pi kL}} (T_1 - T_2) \qquad\qquad \rightarrow \qquad (1.27)$$

$$Q = \frac{\Delta T_{\text{overall}}}{R}$$

where,

$$R = \frac{\ln \left(r_2 / r_1\right)}{2\pi kL} , \; \Delta T = T_1 - T_2$$

1.1.7. One Dimensional Heat Conduction through Hollow Sphere

Step 1

Consider a hollow sphere of inner radius r_1, outer radius r_2, inner temperature T_1, outer temperature T_2 and thermal conductivity k. Let us consider a small element area of thickness 'dr'.

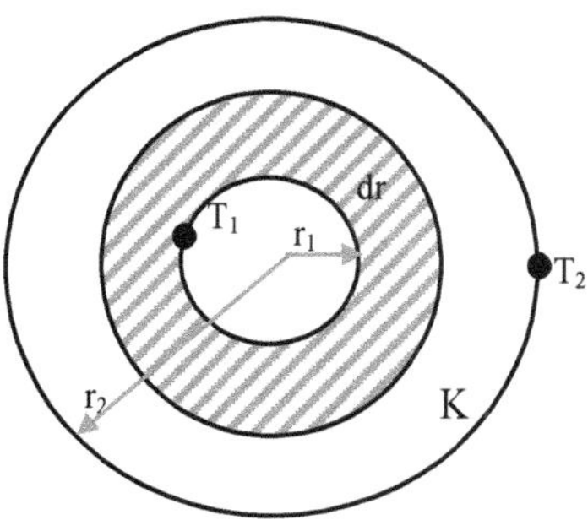

Step 2

3D heat conduction equation through sphere is given by,

$$\frac{d}{dr}\left(r^2 \frac{dT}{dr}\right) = 0 \qquad \rightarrow \qquad (1.28)$$

Step 3

Integrate on both sides in eqn 1.28

$$\int \frac{d}{dr}\; r^2.\; \frac{dT}{dr} = \int 0$$

$$r^2.\frac{dT}{dr} = C_1$$

Again integrate on both sides

$$dT = C_1 \times \frac{dr}{r^2}$$

$$\int dT = C_1 \int \frac{dr}{r^2}$$

$$T = C_1\,(-1/r) + C_2 \qquad \rightarrow \qquad (1.29)$$

Step 4

Boundary Condition (1)	Boundary Condition (2)
$T = T_1$	$T = T_2$
$r = r_1$	$r = r_2$

Applying boundary condition (1) and (2) in eqn (1.29)

$$T_1 = -\frac{C_1}{r_1} + C_2 \qquad \rightarrow \qquad (1.30)$$

$$T_2 = \frac{C_1}{r_2} + C_2 \qquad\qquad \rightarrow \qquad (1.31)$$

Equation (1.30) – (1.31)

$$T_1 - T_2 = -\frac{C_1}{r_1} + C_2) - (-\frac{C_1}{r_2} + C_2)$$

$$T_1 - T_2 = -\frac{C_1}{r_1} + C_2 + \frac{C_1}{r_2} - C_2$$

$$T_1 - T_2 = -\frac{C_1}{r_1} + \frac{C_1}{r_2}$$

$$T_1 - T_2 = C_1 \left(\frac{1}{r_2} - \frac{1}{r_1} \right)$$

$$C_1 = \frac{T_1 - T_2}{\left(\dfrac{1}{r_2} - \dfrac{1}{r_1} \right)}$$

C_1 value substituting in eqn (1.29)

$$T_1 = -\frac{C_1}{r_1} + C_2 \qquad \Rightarrow \qquad C_2 = T_1 + \frac{C_1}{r_1}$$

$$C_2 = T_1 + \frac{T_1 - T_2}{\left(\dfrac{1}{r_2} - \dfrac{1}{r_1} \right)} \cdot \frac{1}{r_1}$$

Step 5

Substituting value of C_1 and C_2 in eqn (1.29)

$$T = -\frac{T_1 - T_2}{\left(\dfrac{1}{r_2} - \dfrac{1}{r_1} \right)} \cdot \frac{1}{r} + \left[T_1 + \frac{T_1 - T_2}{\left(\dfrac{1}{r_2} - \dfrac{1}{r_1} \right)} \cdot \frac{1}{r} \right] \qquad \rightarrow \qquad (1.32)$$

Step 6

From Fourier law of heat conduction

$$Q = -kA \frac{dT}{dr}$$

$$Q = -k \, (4\pi r^2) \cdot \frac{dT}{dr}$$

$$Q. \frac{dr}{r^2} = -4\pi k . dT$$

Both sides integrate

$$Q\int_{r_1}^{r_2} \frac{dr}{r^2} = -4\pi k \int_{T_1}^{T_2} dT$$

$$Q\left(-\frac{1}{r}\right)_{r_1}^{r_2} = -4\pi k [T]_{T_1}^{T_2}$$

$$Q\left[-\frac{1}{r_2} + \frac{1}{r_1}\right] = -4\pi k [T_2 - T_1]$$

$$Q\left[\frac{1}{r_1} - \frac{1}{r_2}\right] = 4\pi k [T_1 - T_2] \qquad \Rightarrow Q\left[\frac{r_2 - r_1}{r_1 r_2}\right] = 4\pi k [T_1 - T_2]$$

$$Q = \frac{4\pi k (T_1 - T_2)}{\dfrac{r_2 - r_1}{r_1 r_2}}$$

$$Q = \frac{T_1 - T_2}{\dfrac{(r_2 - r_1)}{4\pi k (r_1 r_2)}} \qquad\qquad\qquad \rightarrow \qquad (1.33)$$

$$Q = \frac{\Delta T_{overall}}{R}$$

where,

$$\Delta T = T_1 - T_2$$

$$R = \frac{(r_2 - r_1)}{4\pi k (r_1 r_2)}$$

1.1.8. *Newton's Law of Cooling*

Heat transfer by convection is given by newtons law of cooling

$$Q = hA[T_s - T_\infty] \qquad\qquad\qquad \rightarrow \qquad (1.34)$$

where,

A - Area exposed to heat transfer in m^2

h – Heat transfer co-efficient in $w/m^2 k$

T_s – Temperature of the surface in k

T_∞ - Temperature of the fluid in k

1.1.9. *Heat Transfer through a Composite Plane Wall with Inside and Outside Convection*

Step 1

Consider a composite wall of thickness L_1, L_2, L_3 having thermal conductivity k_1, k_2 and k_3 respectively. It is assumed that the interior and exterior surface of the system are subjected to convection at mean temperature T_a and T_b with heat transfer co-efficient h_a and h_b respectively with in the composite wall, the slabs are subjected to conduction.

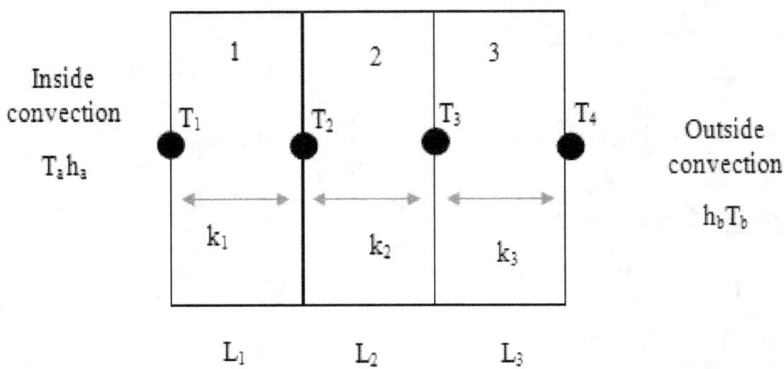

Step 2

From newton's law of cooling

Heat transfer by convection at side A is

$$Q = h_a A\,[T_a - T_1]$$

Heat transfer by conduction at slab

$$(1) \quad Q = \frac{k_1\,A[T_1 - T_2]}{L_1}$$

$$(2) \quad Q = \frac{k_2\,A[T_2 - T_3]}{L_2}$$

$$(3) \quad Q = \frac{k_3\,A[T_3 - T_4]}{L_3}$$

Heat transfer by convection at side B is

$$Q = h_b A[T_4 - T_b]$$

Step 3

Rearrange above equations

$$T_a - T_1 = Q \times \frac{1}{h_a A}$$

$$T_1 - T_2 = Q \times \frac{L1}{k_1 A}$$

$$T_2 - T_3 = Q \times \frac{L_2}{k_2 A}$$

$$T_3 - T_4 = Q \times \frac{L_3}{k_3 A}$$

$$T_4 - T_b = Q \times \frac{L_1}{h_b A}$$

Adding both sides of the above equations

$$T_a - T_b = Q \left[\frac{1}{h_a A} + \frac{L_1}{k_1 A} + \frac{L_2}{k_2 A} + \frac{L_3}{k_3 A} + \frac{1}{h_b A} \right]$$

$$Q = \frac{T_a - T_b}{\dfrac{1}{A}\left[\dfrac{1}{h_a} + \dfrac{L_1}{k_1} + \dfrac{L_2}{k_2} + \dfrac{L_3}{k_3} + \dfrac{1}{h_b} \right]}$$

$$Q = \frac{\Delta T_{overall}}{R}$$

where,

$$\Delta T = T_a - T_b$$

$$R = \frac{1}{A}\left[\frac{1}{h_a} + \frac{L_1}{k_1} + \frac{L_2}{k_2} + \frac{L_3}{k_3} + \frac{1}{h_b} \right]$$

we know that,

$$R = \frac{1}{UA}$$

$$Q = \frac{T_a - T_b}{1/UA}$$

$$Q = UA\,(T_a - T_b)$$

where, υ – overall heat transfer co-efficient (w/m^{2}k)

1.1.10. *Heat Transfer through Composite Pipes or Cylinder with Inside and Outside Convection*

Step 1

A hot fluid at a temperature T_a, with heat transfer co-efficient h_a, flowing through a pipe is separated by two layers from atmosphere. Let the thermal conductivities be k_1 and k_2 on the outside surface heat is being transferred to a cold fluid at a temperature T_b with heat transfer co-efficient h_b.

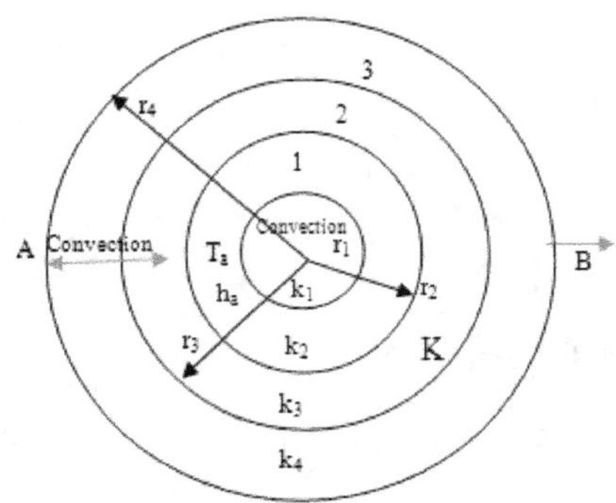

Step 2

Heat transfer by convection at side A is

$Q = h_a\ A\ [T_a - T_1]$

Where area, $A\ = 2\pi r_1 L$

$Q = 2\pi r_1\ L\ h_a\ [T_a\text{-}T_1]$

Heat transfer by conduction at section

$$(1)\quad Q = \frac{2\pi L k_1\ [T_1 - T_2]}{\ln\left(\dfrac{r_2}{r_1}\right)}$$

$$(2)\quad Q = \frac{2\pi L k_2\ [T_2 - T_3]}{\ln\left(\dfrac{r_3}{r_2}\right)}$$

Heat transfer by convection at side B is

$Q = h_b\ A\ [T_3\text{-}T_b]$

$Q = 2\pi r_3 L\ h_b\ [T_3\text{-}T_b]$

Step 3

we know that

$$T_a - T_1 = \frac{Q}{2\pi L r_1 h_a}$$

$$T_1 - T_2 = \frac{Q}{2\pi L\, k_1} \ln\left(\frac{r_2}{r_1}\right)$$

$$T_2 - T_3 = \frac{Q}{2\pi L\, k_2} \ln\left(\frac{r_3}{r_2}\right)$$

$$T_3 - T_b = \frac{Q}{2\pi L\, r_3\, h_b}$$

Adding both sides of the above equations

$$T_a - T_b = \frac{Q}{2\pi L}\left[\frac{1}{h_a\, r_1} + \frac{\ln(r_2/r_1)}{k_1} + \frac{\ln(r_3/r_2)}{k_2} + \frac{1}{h_b\, r_3}\right]$$

$$Q = \frac{2\pi L[T_a - T_b]}{\dfrac{1}{h_a\, r_1} + \dfrac{\ln(r_2/r_1)}{k_1} + \dfrac{\ln(r_3/r_2)}{k_2} + \dfrac{1}{h_b r_3}}$$

$$Q = \frac{[T_a - T_b]}{\dfrac{1}{2\pi L} + \left[\dfrac{1}{h_a r_1} + \dfrac{\ln(r_2/r_1)}{k_1} + \dfrac{\ln(r_3/r_2)}{k_2} + \dfrac{1}{h_b r_3}\right]} \qquad\qquad \rightarrow \qquad (1.35)$$

$$Q = \frac{\Delta T_{overall}}{R}$$

where, $\Delta T_{overall} = T_a - T_b$

$$R = \frac{1}{2\pi L}\left[\frac{1}{h_a\, r_1} + \frac{1}{k_1}\ln\left(\frac{r_2}{r_1}\right) + \frac{1}{k_2}\ln\left(\frac{r_3}{r_2}\right) + \frac{1}{h_b r_3}\right]$$

We known that

$$R = \frac{1}{UA}$$

$$Q = \frac{T_a - T_b}{1/UA}$$

$$Q = UA[T_a - T_b] \qquad\qquad \rightarrow \qquad (1.36)$$

where, U = overall heat transfer co-efficient, $w/m^2 k$

A = Area = $2\pi r_3 L$

			HMT DB (Pg.NO) (8th edition)
S.NO	**DESCRIPTION**	**FORMULA**	
1	Heat transfer through plane wall $$q = \Delta T \frac{overall}{R}$$ 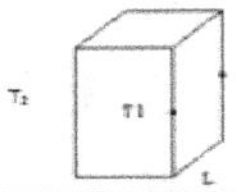	$$Q = \frac{\Delta T}{R}$$ $$\Delta T = (T_1 - T_2)$$ $$R = \frac{L}{KA}$$	Page no – 44
2	Heat transfer is composite of three layer 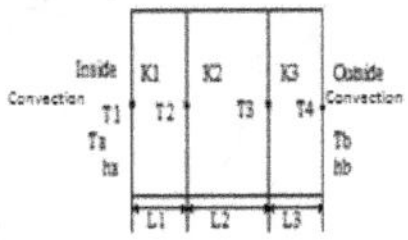 ,	$$Q = \frac{\Delta T}{R} \text{ overall}$$ $$R = \frac{1}{A\left[\dfrac{1}{ha} + \dfrac{L1}{K1} + \dfrac{L2}{K2} + \dfrac{L3}{K3} + \dfrac{1}{hb}\right]}$$ $$\Delta T = (T_a - T_b)$$ $$Q = \frac{Ta-Tb}{R} = \frac{Ta-T1}{R_a} = \frac{T1-T2}{R_1} = \frac{T2-T3}{R_2} = \frac{T3-T4}{R_3} = \frac{T4-Tb}{R_b}$$ (Interference (temperature relations) $$Q = \frac{T_1 - T_2}{1/UA} , \quad Q = UA[T_1 - T_2]$$	Page No – 45
3	Heat transfer is composed with three layer without ha, hb 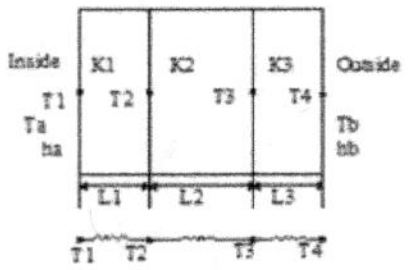	$$Q = \frac{\Delta T}{R} \text{ overall}$$ $$R = \frac{1}{A\left[\dfrac{L1}{K1} + \dfrac{L2}{K2} + \dfrac{L3}{K3}\right]}$$ $$(h_a , h_b = 0)$$ $$\Delta T = (T_1 - T_4)$$ $$Q = \frac{T_1 - T_4}{R} = \frac{T_1 - T_2}{R_1} = \frac{T_2 - T_3}{R_2} = \frac{T_3 - T_4}{R_3}$$	Page No – 45
4	Convection heat transfer for newton's law of woling	$$Q = hA[Ts - T\infty]$$	Page No – 149
5	Series – parallel composite layer 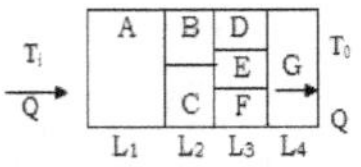	$$R = R_1 + R_2 + R_3 + R_4$$	Page No – 47
	Where,	Q = heat transfer in walls h_a, h_b = heat transfer coefficients in $\frac{W}{m^2}.K$ Δ = over all difference in temp in ^0c (or) K R = Thermal resistance of slab (K/W) K = Thermal conductivity of material (W/m.k) U = Overall heat transfer co-efficient (w/m^{2}k) L = Thickness (m)	

Table 1.1: Required Formula for Slab

1.1.11. Solved Problems on Slabs

Compute the heat loss per square meter surface area of a 40cm thick furnace wall having surface temperature of 300°C and 50°C. If the thermal conductivity of the wall material is given by K = 0.005T – 5 x $10^{-6}T^2$ where T = temperature in °C.

Given

L = 40cm = 0.4cm

T_1 = 300°C

T_2 = 50°C

k = 0.005T – 5 x $10^{-6}T^2$

To Find

Heat transfer rate per square meter surface area

$$\frac{Q}{A} = ?$$

Solution

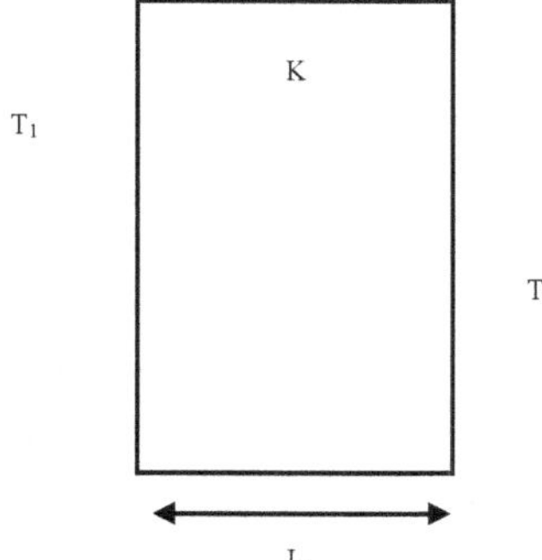

Step 1

we know that,

$$Q = \frac{\Delta T}{R} = \frac{T_1 - T_2}{L/kA}$$

$$\frac{Q}{A} = \frac{T_1 - T_2}{L/k}$$

Step 2

$$k = 0.005T - 5 \times 10^{-6} \, T^2$$

$$T = \frac{T_1 + T_2}{2}$$

$$T = \frac{300 + 50}{2} = 175\,^\circ C$$

$$k = 0.005 \times 175 - 5 \times 10^{-6} \, (175)^2$$

$$k = 0.7218 \text{ w/m}^\circ C$$

Step 3

$$\frac{Q}{A} = \frac{300 - 50}{0.4 / 0.7218}$$

$$\frac{Q}{A} = \frac{300 - 50}{0.5542} \qquad = \qquad 451.1006 \text{ w/m}^2$$

Result

$$Q/A \qquad = \qquad 451.1006 \text{ w/m}^2$$

(Heat transfer, per m^2)

1.1.12. Solved Problems on Composite Plane Wall

1. A furnace wall is made up of three layer of thickness 25cm, 10cm and 15cm with thermal conductivities of 1.65, k and 9.2 w/mk respectively. The inside is exposed to gases at 1250°C with a convection co-efficient of 25w/m²k and the inside surface is at 1100°C, the outside surface is exposed to air at 25°C with convection co-efficient of 12w/m²k. Determine (i) the unknown thermal conductivity, (ii) the overall heat transfer co-efficient (iii) All the surface temperature (iv) overall thermal resistance (v) The rate of heat transfer

Given Data

Thickness, L_1	= 25cm	=	0.25m
L_2	= 10cm	=	0.10m
L_3	= 15cm	=	0.15cm

Thermal conductivity,

k_1 = 1.65 w/mk

k_2 = kw/mk

$k_3 = 9.2 \ w/mk$

Inside gas temperature, $T_a = 1250°C + 273 = 1523 \ k$

Outside air temperature, $T_b = 25°C + 273 = 298k$

Inside heat transfer coefficient, $h_a = 25 \ w/m^2k$

Outside heat transfer coefficient, $h_b = 12 \ w/m^2k$

Inner surface temperature, $T_1 = 1100°C + 273 = 1373k$

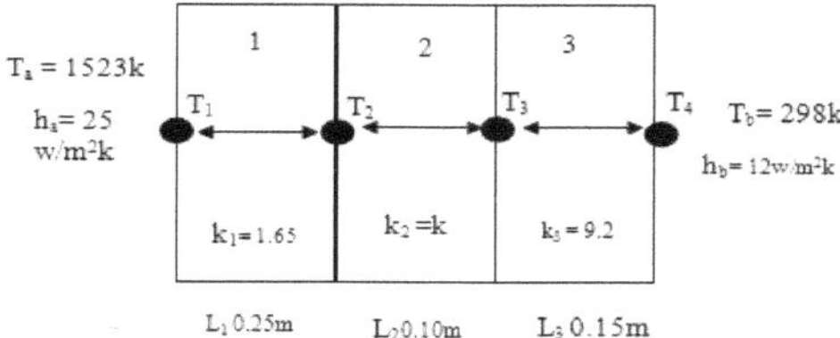

To Find

(i) Unknown thermal conductivity, k_2

(ii) Overall heat transfer coefficient, U

(iii) surface temperature T_2, T_3, T_4

(iv) Overall thermal resistance, R

Solution

Step 1

heat transfer rate

we know that,

Heat transfer $Q = h_a A \ (T_a - T_1)$

$$\Rightarrow \quad \frac{Q}{A} = h_a \ (T_a - T_1) = 25 \ (1523 - 1373)$$

$$\frac{Q}{A} = 3750 w/m^2$$

Step 2

Thermal conductivity k_2,

Heat flow, $Q = \dfrac{\Delta T_{overall}}{R}$ [From HMT D.B.Pg. No. 45]

$$Q = \frac{T_a - T_b}{\dfrac{1}{A}\left[\dfrac{1}{h_a} + \dfrac{L_1}{k_1} + \dfrac{L_2}{k_2} + \dfrac{L_3}{k_3} + \dfrac{1}{h_b}\right]}$$

$$\frac{Q}{A} = \frac{T_a - T_b}{\dfrac{1}{h_a} + \dfrac{L_1}{k_1} + \dfrac{L_2}{k_2} + \dfrac{L_3}{k_3} + \dfrac{1}{h_b}}$$

$$3750 = \frac{1523 - 298}{\dfrac{1}{25} + \dfrac{0.25}{1.65} + \dfrac{0.10}{k_2} + \dfrac{0.15}{9.2} + \dfrac{1}{12}}$$

$$3750 = \frac{1225}{0.2911 + \dfrac{0.10}{k_2}}$$

$$3750\left(0.2911 + \frac{0.10}{k_2}\right) = 1225$$

$$0.2911 + \frac{0.10}{k_2} = 0.3266$$

Thermal conductivity, k_2 = 2.816 w/mk

Step 3

Overall thermal resistance, R

$$R_{total} = \frac{1}{A}\left[\frac{1}{h_a} + \frac{L_1}{k_1} + \frac{L_2}{k_2} + \frac{L_3}{k_3} + \frac{1}{h_b}\right]$$

For unit area [Take A = 1 m²]

$$R_{total} = \frac{1}{25} + \frac{0.25}{1.65} + \frac{0.1}{2.816} + \frac{0.15}{9.2} + \frac{1}{12}$$

R_{total} = 0.3267 w/m²

Step 4

Overall heat transfer co – efficient, U

$Q = UA (T_a - T_b)$

$$\frac{Q}{A} = U (T_a - T_b)$$

$3750 = U (1523 - 298)$

U = 3.061 w/m²k

Alter method

$$U = \frac{1}{R_{total}}$$

$$= \frac{1}{0.3267}$$

U = 3.06 w/m²k

Step 5

1. Surface temperatures T_2, T_3, T_4

   We know that,

$$Q = \frac{T_a - T_b}{R} = \frac{T_a - T_1}{R_a} = \frac{T_1 - T_2}{R_1} = \frac{T_2 - T_3}{R_2} = \frac{T_3 - T_4}{R_3} = \frac{T_4 - T_b}{R_b} \quad \ldots\ldots(1)$$

(1) $\Rightarrow Q = \dfrac{T_1 - T_2}{R_1}$ \qquad Where $R_1 = \dfrac{L_1}{k_1 A}$

$$\Rightarrow \frac{Q}{A} = \frac{T_1 - T_2}{\dfrac{L_1}{k_1}}$$

$$\Rightarrow 3750 = \frac{1373 - T_2}{\dfrac{0.25}{1.65}}$$

$$\Rightarrow T_2 = 804.8 \text{ k}$$

(1) $\Rightarrow \qquad Q = \dfrac{T_2 - T_3}{R_2}$ \quad Where $R_2 = \dfrac{L_2}{k_2 A}$

$$\frac{Q}{A} = \frac{T_2 - T_3}{\dfrac{L_2}{k_2}}$$

$$3750 = \frac{804.8 - T_3}{\dfrac{0.10}{2.816}}$$

$$\Rightarrow T_3 = 671.45 \text{ k}$$

(1) $\Rightarrow Q = \dfrac{T_3 - T_4}{R_3}$ \quad Where $R_3 = \dfrac{L_3}{k_3 A}$

$$\frac{Q}{A} = \frac{T_3 - T_4}{\dfrac{L_3}{k_3}}$$

$$3750 = \frac{671.45 - T_4}{\dfrac{0.15}{9.2}}$$

$$T_4 = 610.30 \text{ k}.$$

Result

    (i) Unknown thermal conductivity, k_2 = 2.816 w/mk

    (ii) The overall heat transfer co-efficient, U = 3.06 w/m²k

    (iii) All the surface temperatures, T_2 = 804.8 k

        T_3 = 671.45 k

        T_4 = 610.30 k

    (iv) Over all thermal resistance, R_{total}= 0.3267 w/m²

    (v) The rate of heat transfer, $\dfrac{Q}{A}$ = 3750 w/m²

2.   A furnace wall consist of 200mm layer of refractory bricks, 6mm layer of steel plate and a 100 mm layer of insulation bricks on the furnace side maximum temperature is 1150°C and minimum temperature is 40°C on the outer side of the wall. The heat loss from the wall is 400 w/m². There is thin layer of air between the refractory brick and steel plate. Thermal conductivity for the three layers are 1.52, 45 and 0.138 w/m°c respectively. Find:

    1.   To how many mm of insulation brick is the air layer equivalent.

    2.   What is the temperature of the outer surface of the steel plates?

Given

Refractory bricks	L_1 = 200 mm = 0.2 m,	k_1 = 1.52 w/m°C
Air	L_2 = ?	k_2 = ?
Steel Plate	L_3 = 6 mm = 0.06 m,	k_3 = 45 w/m°C
Insulation bricks	L_4 = 100 mm = 0.1 m,	k_4 = 0.138 w/m°C

T_1 = 1150°C + 273 = 1423 k

T_5 = 40°C + 273 = 313 k

Heat loss $\dfrac{Q}{A}$ = 400 w/m²

To Find

    (i) Air gap equivalent to L_2 mm of insulation brick.

    (ii) Temperature on the outer surface of steel plate (T_4)

Step 1

Thermal conductivity of air layer k_2, k_2 = 0.062 w/m°C [From HMT data book pg.No – 34]

(k_2 value obtain from temperature T)

$$T = \frac{T_1 + T_5}{2} = \frac{1150 + 40}{2} = 595°C \simeq 600°C$$

Step 2

(i) Air Layer Equivalent L₂,

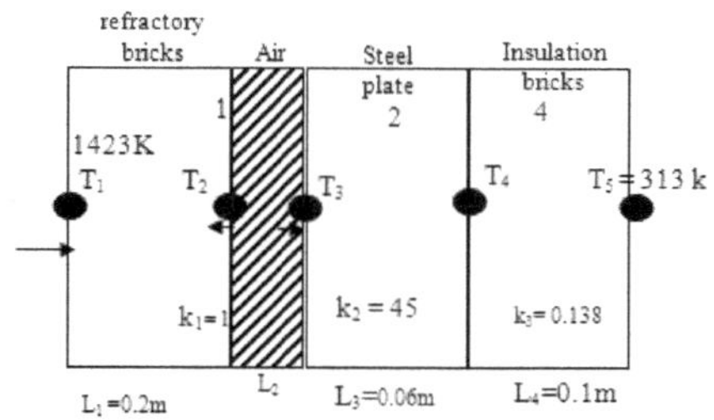

We know that,

Heat loss

$$Q = \frac{\Delta T_{overall}}{R}$$

$$Q = \frac{T_1 - T_5}{\frac{1}{A}\left[\frac{1}{h_a} + \frac{L_1}{k_1} + \frac{L_2}{k_2} + \frac{L_3}{k_3} + \frac{L_4}{k_4} + \frac{1}{h_b}\right]}$$

Now given h_a, h_b, so,

$$\frac{Q}{A} = \frac{T_1 - T_5}{\frac{L_1}{k_1} + \frac{L_2}{k_2} + \frac{L_3}{k_3} + \frac{L_4}{k_4}}$$

$$400 = \frac{1150 - 40}{\frac{0.2}{1.52} + \frac{L_2}{0.062} + \frac{0.006}{45} + \frac{0.1}{0.138}}$$

$$400 = \frac{1150 - 40}{0.8581 + \frac{L_2}{0.062}}$$

$$0.8581 + \frac{L_2}{0.062} = 2.775$$

$$\frac{L_2}{0.062} = 1.9169$$

$L_2 = 0.1189$ m

$L_2 = 118.9$ mm

Thus, 118.9 mm of insulation brick is the equivalent to air gap layer.

Step 3

(ii) Temperature on Outer Surface of Steel Plates

$$Q = \frac{T_1 - T_2}{R_1} = \frac{T_2 - T_3}{R_2} = \frac{T_3 - T_4}{R_3} = \frac{T_4 - T_5}{R_4} \qquad \text{.......(1)}$$

$$(1) \Rightarrow Q = \frac{T_4 - T_5}{R_4} \qquad \text{Where} \qquad R_4 = \frac{L_4}{k_4 A}$$

$$\frac{Q}{A} = \frac{T_4 - T_5}{\dfrac{L_4}{k_4}}$$

$$400 = \frac{T_4 - 40}{\dfrac{0.1}{0.138}}$$

$$T_4 = 329.86°C.$$

Result

(i)   Air layer equivalent L_2 = 118.9 mm

(ii) Temperature on outer surface of steel plate T_4 = 329.86°C

3. An external wall of a house is made up of 10cm common brick (k = 0.7 w/mk). Followed by a 4 cm layer of gypsum plaster (k = 0.48 w/mk). What thickness of loosly packed insulation (k = 0.065 w/mk) should be added to reduce the next loss through the wall by 80%.

Given

Brick,	L_1	= 10 cm = 0.1 m,	k_1 = 0.7 w/mk
Gypsum,	L_2	= 4 cm = 0.04m,	k_2 = 0.48 w/mk
Insulation,	L_3	= ?,	k_3 = 0.065 w/mk.

To Find

Thickness of insulation L_3.

Solution

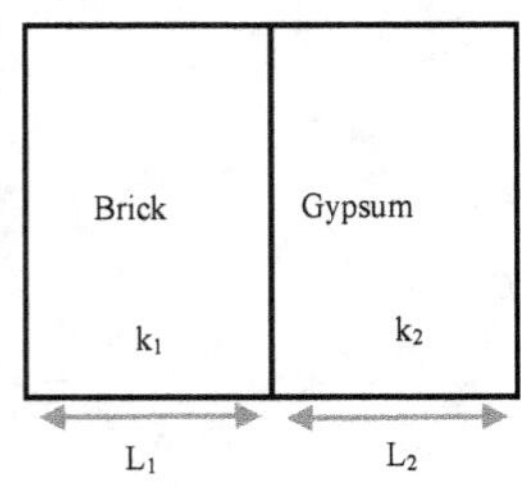

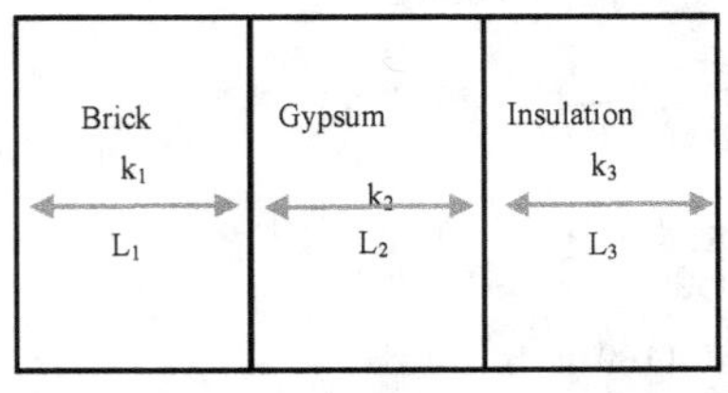

Case(i)

Heat flow rate,

$$Q = \frac{\Delta T_{overall}}{R}$$

$$Q = \frac{\Delta T}{\dfrac{L_1}{k_1 A} + \dfrac{L_2}{k_2 A}} \qquad\qquad [\because A = 1 m^2]\ [\text{HMT D.B. Pg.No.45}]$$

$$100 = \frac{\Delta T}{\dfrac{0.1}{0.7} + \dfrac{0.04}{0.48}}$$

$\Delta T = 22.61$ k.

$$Q = \frac{\Delta T}{R}$$

$$Q = \frac{\Delta T}{\dfrac{L_1}{k_1 A} + \dfrac{L_2}{k_2 A} + \dfrac{L_3}{k_3 A}}$$

Heat loss is reduced 80%

Due to insulation, so

$$20 = \frac{22.61}{\dfrac{0.1}{0.7} + \dfrac{0.04}{0.48} + \dfrac{L_3}{0.065}}$$

$L_3 = 0.0587$ m

Result

Thickness of insulation, $L_3 = 0.0587$ m.

4. Find the heat flow rate through the composite wall as shown in fig.5. Assume one dimension flow. Take $k_A = 150$ w/m°C, $k_B = 30$ w/m°C and $k_C = 65$ w/m°C, $k_D = 50$ w/m°C.

Given

$k_A = 150$ w/m°C $\qquad L_A = 3$ cm $= 0.03$ m

$k_B = 30$ w/m°C $\qquad L_B = 8$ cm $= 0.08$ m

$k_C = 65$ w/m°C $\qquad L_C = 8$ cm $= 0.08$ m

$k_D = 50 \text{ w/m}°C$ $\qquad$ $L_D = 5 \text{ cm} = 0.05 \text{ m}$

$T_1 = 400°C$ $\qquad$ $T_4 = 60°C$

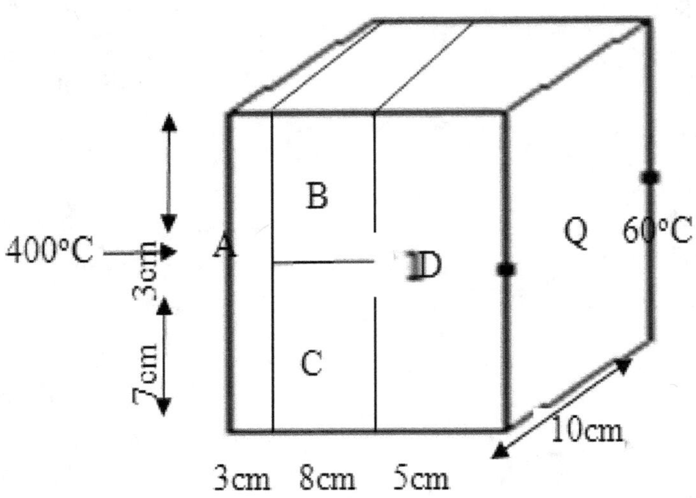

Step 1

Areas normal to heat transfer

$A_A = 0.1 \times 0.1 = 0.01 \text{ m}^2$

$A_B = 0.1 \times 0.03 = 0.003 \text{ m}^2$ $\qquad\qquad$ $[\therefore 0.03 + 0.07 = 0.1 \text{ m}^2]$

$A_C = 0.1 \times 0.07 = 0.007 \text{ m}^2$

$A_D = 0.1 \times 0.1 = 0.01 \text{ m}^2$

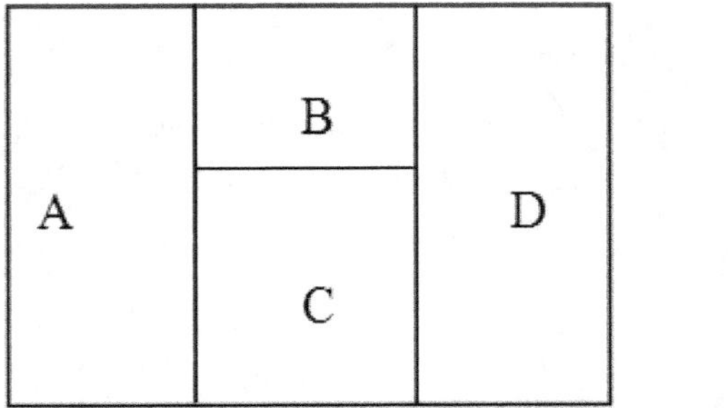

Step 2

Heat flow rate, [From HMT Data book Pg.No . 47] $Q = \dfrac{(\Delta T)_{overall}}{R_{total}}$

$R_{total} = R_1 + R_2 + R_3$

$R_1 = R_A,$ $R_2 = \dfrac{R_B . R_C}{R_B + R_C}$ $R_3 = R_D$

$$R_A = \dfrac{L_A}{k_A \, A_A} = \dfrac{0.03}{150 \times 0.01} = 0.02 \; °C/w$$

$$R_B = \dfrac{L_B}{k_B \, A_B} = \dfrac{0.08}{30 \times 0.003} = 0.8888 \; °C/w$$

$$R_C = \dfrac{L_C}{k_C \, A_C} = \dfrac{0.08}{65 \times 0.007} = 0.17582 \; °C/w$$

$$R_D = \dfrac{L_D}{k_D \, A_D} = \dfrac{0.05}{50 \times 0.01} = 0.1 \; °C/w$$

$R_1 = 0.02°C/w$

$$R_2 = \dfrac{R_B . R_C}{R_B + R_C} = \dfrac{0.8888 \times 0.17582}{0.8888 + 0.17582}$$

$R_2 = 0.14678°c/w$

$R_3 = 0.1 \; °c/w$

$R_{total} = R_1 + R_2 + R_3$

$\quad = 0.02 + 0.14678 + 0.1$

$R_{total} = 0.26675°c/w$

Step 3

$$Q = \dfrac{T_1 - T_4}{R_{total}} \qquad = \dfrac{400 - 60}{0.26675} \qquad = 1274.60 \; w$$

Result

Heat flow rate Q = 1274.60 w.

Table 1.2: Required Formula for Cylinder			
S.NO	**DESCRIPTION**	**FORMULA**	**HMT DBP NO**
1	Heat transfer through a hollow cylinder,	$$Q = \frac{(\Delta T)\,overall}{R}$$ $$\Delta T = (T1 - T2)$$ $$R = \frac{1}{2\pi LK}\ln\left(\frac{r2}{r1}\right)$$	Pg.No – 44 Pg.No – 45
2	Heat transfer through a hollow cylinder with two insulation	$$Q = \left(\frac{\Delta T}{R}\right) over\ all$$ $$\Delta T = (Ta - Tb)$$ $$R = \frac{1}{2\pi L\left[\frac{1}{h_a r1} + \frac{1}{K1}\ln\left(\frac{r2}{r1}\right) + \frac{1}{K2}\ln\left(\frac{r3}{r2}\right) + \frac{1}{K3}\ln\left(\frac{r4}{r3}\right) + \frac{1}{h_b r4}\right]}$$ $$Q = \frac{T_a - T_b}{R} = \frac{T_a - T_1}{R_a} = \frac{T_1 - T_2}{R_1} = \frac{T_2 - T_3}{R_2} = \frac{T_3 - T_4}{R_3} = \frac{T_4 - T_b}{R_4}$$	Pg.No – 46

Where,

L	=	Length in M	
Q	=	heat transfer in walls	
ΔT	=	Temperature change in a layer in K (or) $^{\circ}$C	
R	=	Thermal resistance in k/w	
r	=	Radius in M.	
Q	=	UA (Ta – Tb)	
U	=	Overall heat transfer co-effective w/m^{2}k	
A	=	Area (2π r$_4$L)m^2	

1.1.13. Solved Problems of Cylinder

1. A hollow cylinder 5 cm inner radius and 10 cm outer radius has inner surface temperature of 200°C and outer surface temperature of 100°C. if the thermal conductivity is 70 w/mk. Find heat transfer per unit length.

Given

Inner radius, r_1 = 5 cm = 0.05 m

Outer radius, r_2 = 10 cm = 0.1 m

Inner surface temperature T_1 = 200°C + 273 = 473 k

Outer surface temperature T_2 = 100°C + 273 = 373 k

Thermal conductivity, k = 70 w/mk.

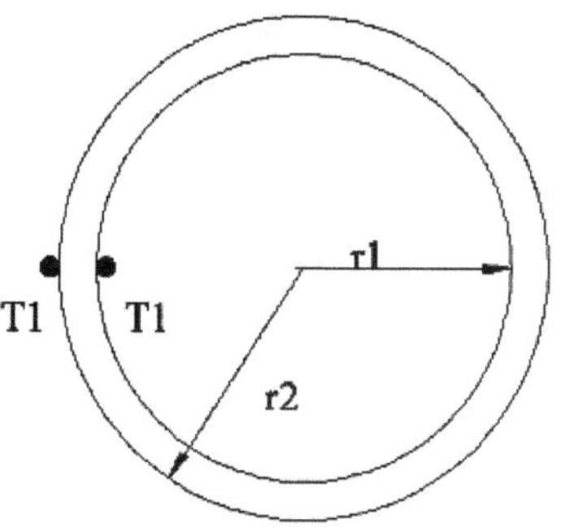

To Find

Heat flow per unit length Q/L

Solution

Step 1

Heat transfer through hallow cylinder is given by

$$Q = \frac{\Delta T_{overall}}{R}$$

[From HMT Data book Pg.No- 45]

Where $\Delta T = T_1 - T_2$

$$R = \frac{1}{2\pi Lk} \, ln\left(\frac{r_2}{r_1}\right)$$

$$\Rightarrow Q = \frac{T_1 - T_2}{\dfrac{1}{2\pi Lk} ln\left(\dfrac{r_2}{r_1}\right)}$$

$$\Rightarrow Q = \frac{2\pi Lk \,\, (T_1 - T_2)}{ln\left(\dfrac{r_2}{r_1}\right)}$$

$$\Rightarrow \frac{Q}{L} = \frac{2\pi \times 70 \, (473 - 373)}{ln\left(\dfrac{0.1}{0.05}\right)}$$

Q/L = 63453.04 w/m = 63.453 kw/m

Result

Heat transfer per unit length, Q/L = 63.453 kw/m.

2. A steel tube 5 cm ID, 7.6 cm OD and k = 15 w/m°c is covered with an insulative covering of thickness 2 cm and k = 0.2 w/m°C. A hot gas at 330°C with h = 400 w/m²°C flows inside the tube. The outer surface of the insulation is exposed to cooler air at 30°C with h = 60 w/m²°C calculate the heat loss from the tube to the air for 10 m of the tube and temperature drops resulting from the thermal resistances of the hot gas flow. The steel tube, the insulation layer and the outside air. Also calculate contact temperatures and overall heat transfer co – efficient.

Given

Inner diameter of Steel, d_1 = 5 cm = 0.05m.

Inner radius, r_1 = 0.025 m

Outer diameter of steel, d_2 = 7.6 cm = 0.076 m

Outer radius, r_2 = 0.038m

Radius, $r_3 = r_2$ + thickness of insulation

$\qquad$ = 0.038 + 0.02 m

r_3 = 0.058m

Thermal conductivity of steel, k_1 = 15 w/m°C

Thermal conductivity of insulation, k_2 = 0.2 w/m°C

Hot gas temperature, T_a = 330°C + 273 = 603 k

Heat transfer co-efficient at inner side, h_a = 400 w/m²°C

Ambient air temperature, T_b = 30°C + 273 = 303 k

Heat transfer co–efficient at outer side, h_b = 60 w/m²C

Length, L = 10 m

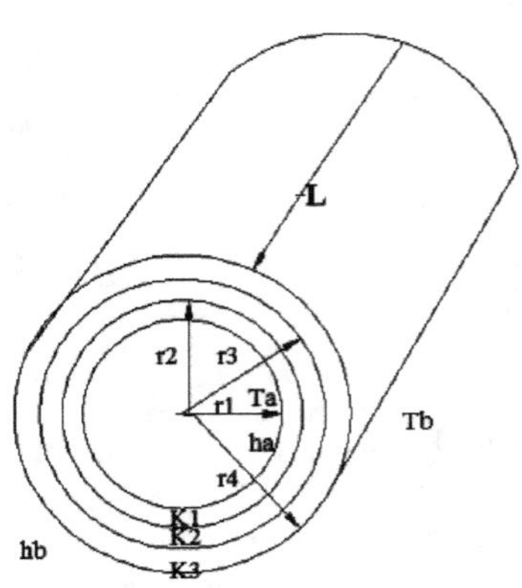

To Find

(i) Heat loss Q

(ii) Temperature drops : $(T_a - T_1)$, $(T_1 - T_2)$, $(T_2 - T_3)$ and $(T_3 - T_b)$

(iii) Interface temperatures : T_2

(iv) Overall heat transfer co-efficient, U

Solution

Step 1

heat flow

$$Q = \frac{\Delta T_{overall}}{R} \qquad \text{[From HMT Data Book Pg.No 46 and 47]}$$

$$\Rightarrow Q = \frac{T_a - T_b}{\dfrac{1}{2\pi L}\left[\dfrac{1}{h_a r_1} + \dfrac{1}{k_1}\ln\left(\dfrac{r_2}{r_1}\right) + \dfrac{1}{k_2}\ln\left(\dfrac{r_3}{r_2}\right) + \dfrac{1}{k_3}\ln\left(\dfrac{r_4}{r_3}\right) + \dfrac{1}{h_b r_4}\right]}$$

[The terms k_3 and r_4 are not given, so neglect that terms]

$$\Rightarrow Q = \frac{T_a - T_b}{\dfrac{1}{2\pi L}\left[\dfrac{1}{h_a r_1} + \dfrac{1}{k_1}\ln\left(\dfrac{r_2}{r_1}\right) + \dfrac{1}{k_2}\ln\left(\dfrac{r_3}{r_2}\right) + \dfrac{1}{h_b r_3}\right]}$$

$$\Rightarrow Q = \frac{603 - 303}{\dfrac{1}{2\pi \times 10}\left[\dfrac{1}{400 \times 0.025} + \dfrac{1}{15}\ln\left(\dfrac{0.038}{0.025}\right) + \dfrac{1}{0.2}\ln\left(\dfrac{0.058}{0.038}\right) + \dfrac{1}{60 \times 0.058}\right]}$$

$$\Rightarrow Q = 7451.72 \text{ w.}$$

Step 2

Interface Temperatures,

$$Q = \frac{T_a - T_b}{R} = \frac{T_a - T_1}{R_a} = \frac{T_1 - T_2}{R_1} = \frac{T_2 - T_3}{R_2} = \frac{T_3 - T_b}{R_b} \qquad \rightarrow (1)$$

$$(1) \Rightarrow Q = \frac{T_a - T_1}{R_a} \qquad\qquad \therefore R_a = \frac{1}{2\pi L}\left[\dfrac{1}{h_a r_1}\right]$$

$$Q = \frac{T_a - T_1}{\dfrac{1}{2\pi L} \times \dfrac{1}{h_a r_1}}$$

$$7451.72 = \frac{T_a - T_1}{\dfrac{1}{2\pi \times 10} \times \left[\dfrac{1}{400 \times 0.025}\right]}$$

$\Rightarrow T_a - T_1 = 11.859 \text{ k}$

Temperature drop across hot gas flow, $T_a - T_1 = 11.859 \text{ k}$

$(1) \Rightarrow Q = \dfrac{T_1 - T_2}{R_1}$
$\qquad\qquad \therefore R_1 = \dfrac{1}{2\pi L}\left[\dfrac{1}{k_1} ln\left(\dfrac{r_2}{r_1}\right)\right]$

$$= \dfrac{T_1 - T_2}{\dfrac{1}{2\pi \times 10}\left[\dfrac{1}{k_1} ln\left(\dfrac{r_2}{r_1}\right)\right]}$$

$$7451.72 = \dfrac{T_1 - T_2}{\dfrac{1}{2\pi \times 10}\left[\dfrac{1}{15} ln\left[\dfrac{0.038}{0.025}\right]\right]}$$

$\Rightarrow T_1 - T_2 = 3.310 \text{ k}$

Temperature drop across the steel tube, $T_1 - T_2 = 3.310 \text{ k}$

$(1) \Rightarrow Q = \dfrac{T_2 - T_3}{R_2}$
$\qquad\qquad \therefore R_2 = \dfrac{1}{2\pi L}\left[\dfrac{1}{k_2} ln\left(\dfrac{r_3}{r_2}\right)\right]$

$$= \dfrac{T_2 - T_3}{\dfrac{1}{2\pi L}\left[\dfrac{1}{k_2} ln\left(\dfrac{r_3}{r_2}\right)\right]}$$

$$7451.72 = \dfrac{T_2 - T_3}{\dfrac{1}{2\pi \times 10}\left[\dfrac{1}{0.2} ln\left[\dfrac{0.058}{0.038}\right]\right]}$$

$\Rightarrow T_2 - T_3 = 250.75 \text{ k}$

Temperature drop across the insulation, $T_2 - T_3 = 250.74 \text{ k}$

$(1) \Rightarrow Q = \dfrac{T_3 - T_b}{R_b}$

$Q = \dfrac{T_3 - T_b}{\dfrac{1}{2\pi L}\left(\dfrac{1}{h_b r_3}\right)}$
$\qquad\qquad \therefore R_b = \dfrac{1}{2\pi L}\left(\dfrac{1}{h_b r_3}\right)$

$$7451.72 = \dfrac{T_3 - T_b}{\dfrac{1}{2\pi \times 10}\left[\dfrac{1}{60 \times 0.058}\right]}$$

$\Rightarrow T_3 - T_b = 34.07 \text{ k}$

Temperature drop across the out side air, $T_3 - T_b = 34.07 \text{ k}$

Step 3

Contact temperature T_2.

We know that,

$T_a - T_1 = 11.859$ k

$T_1 = 603 - 11.859$

$T_1 = 591.141$ k

$T_1 - T_2 = 3.310$ k

$T_2 = 591.141 - 3.310$

$T_2 = 587.831$ k

$T_2 - T_3 = 250.75$ k

$T_3 = 587.831 - 250.75$

$T_3 = 337.081$ k

Step 4

Overall heat transfer co-efficient, U

$Q = UA\,(T_a - T_b)$

$Q = U \times 2\pi r_3 L \times (T_a - T_b)$ $\therefore A = 2\pi r_3 L$

$\Rightarrow 7451.72 = U \times 2\pi \times 0.058 \times 10 \times (603 - 303)$

$$U = \frac{7451.72}{1092.72}$$

$U = 6.819$ w/m^{2}k

Result

(i)   Heat loss Q = 7451.72 w

(ii)   Temperature drops, $T_a - T_1 = 11.859$ k

                                        $T_1 - T_2 = 3.310$ k

                                        $T_2 - T_3 = 250.75$ k

                                        $T_3 - T_b = 34.07$ k

(iii) Interface temperature, $T_2 = 587.831$ k

(iv) Overall heat transfer co-efficient U = 6.819 w/m^{2}k.

		Table 1.3: Required Formula for Sphere		
S.NO	**DESCRIPTION**	**FORMULA**		**HMT DBP NO**
1	Heat transfer through a sphere,	$Q = \dfrac{(\Delta T)\,overall}{R}$ $\Delta T = (Ta - Tb)\ \ or\ \ (T1 - T2)$ $R = \dfrac{1}{4\pi\left[\dfrac{1}{ha\,r_1^2} + \dfrac{1}{K1\left[\frac{1}{r_1}-\frac{1}{r_2}\right]+\frac{1}{hb\,r_2^2}}\right]}$ $(ha, hb\ is\ not\ given)$ $R = \dfrac{1}{4\pi\left[\dfrac{1}{K1\left[\frac{1}{r_1}-\frac{1}{r_2}\right]}\right]}$ $Q = \dfrac{T1 - T2}{4\pi\left[\dfrac{1}{ha\,r_1^2}+\dfrac{1}{K1\left[\frac{1}{r_1}-\frac{1}{r_2}\right]+\frac{1}{hb\,r_2^2}}\right]}$		Pg.No – 45
2	Heat transfer through a composite sphere	$Q = (\Delta T)/R$ $R = \dfrac{1}{4\pi\left[\dfrac{1}{ha\,r_1^2} + \dfrac{1}{K1\left[\frac{1}{r_1}-\frac{1}{r_2}\right]+\dfrac{1}{K2\left[\frac{1}{r_2}-\frac{1}{r_3}\right]}+1/hb\,r_3^2}\right]}$ $\Delta T = T_a - T_b$		Pg.No – 46
	Where, L – Length in m			
	Q – Heat transfer in walls			
	R – Thermal resistance in k/w			
	r – radius in m			
	ΔT – Temperature change in K (or) ºC			

1.1.15. Solved Problem on Hollow Sphere

1. A Hollow Sphere 1.2m inner diameter and 1.7 m, outer diameter is having a thermal conductivity of 1 w/mk. The inner surface temperature is 70 k and outer surface temperature is 300 k. determine, (i) Heat transfer rate (ii) temperature at a radius of 650 mm.

Given

$d_1 = 1.2$ m

$r_1 = 0.6$ m

$d_2 = 1.7$ m

$r_2 = 0.85$ m

$k_1 = 1$ w/mk

$T_1 = 70$ k

$T_2 = 300$ k

$R = 650$ mm $= 0.65$ m

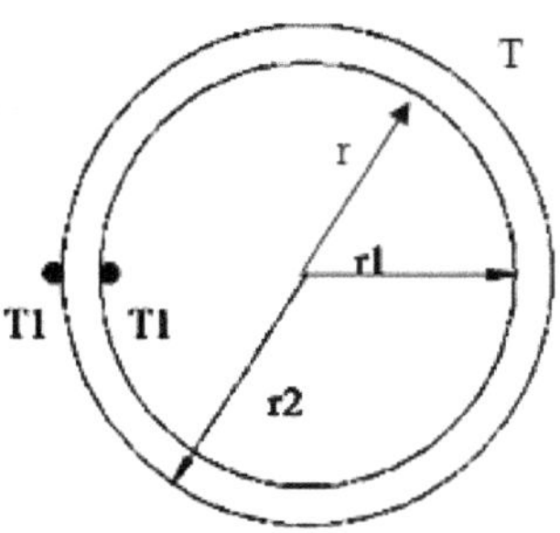

To Find

(i) Heat transfer rate, (Q)

(ii) Temperature at a radius of 650 mm

Solution

Step 1

(i) Heat transfer rate, (Q)

Heat transfer $Q = \dfrac{\Delta T_{overall}}{R}$

[From HMT D.B pg.No 45]

Where $\Delta T = T_a - T_b = T_1 - T_2$

$$R = \frac{1}{4\pi}\left[\frac{1}{h_a r_1^2} + \frac{1}{k_1}\left[\frac{1}{r_1} - \frac{1}{r_2}\right] + \frac{1}{h_b r_2^2}\right]$$

$$Q = \frac{T_1 - T_2}{\dfrac{1}{4\pi}\left[\dfrac{1}{h_a r_1^2} + \dfrac{1}{k_1}\left[\dfrac{1}{r_1} - \dfrac{1}{r_2}\right] + \dfrac{1}{h_b r_2^2}\right]}$$

[The terms h_a and h_b are not given. So, neglect that terms]

$$\Rightarrow Q = \cfrac{T_1 - T_2}{\cfrac{1}{4\pi}\left[\cfrac{1}{k_1}\left[\cfrac{1}{r_1} - \cfrac{1}{r_2}\right]\right]}$$

$$= \cfrac{70 - 300}{\cfrac{1}{4\pi}\left[\cfrac{1}{1}\left[\cfrac{1}{0.6} - \cfrac{1}{0.85}\right]\right]}$$

$$\Rightarrow Q = -5896.14 \text{ w}$$

[The negative sign indicates that heat flows from outside to inside]

Step 2

(ii) Temperature at a radius = r = 0.65 m

Put $T_2 = T$ and $r_2 = r$

$$Q = \cfrac{T_1 - T}{\cfrac{1}{4\pi}\left[\cfrac{1}{k_1}\left[\cfrac{1}{r_1} - \cfrac{1}{r}\right]\right]}$$

$$-5896.14 = \cfrac{70 - T}{\cfrac{1}{4\pi}\left[\cfrac{1}{1}\left[\cfrac{1}{0.6} - \cfrac{1}{0.65}\right]\right]}$$

T = 130.15 k.

Result

(i)   Heat transfer rate, Q = -5896.14 w

(ii) Temperature at a radius of 650 mm = 130.15 k.

2. A hollow sphere (k = 65 w/mk) of 120 mm inner diameter and 350 mm outer diameter is covered 10mm layer of insulation (k = 10 w/mk) the inside and outside temperatures are 500°C and 50°C respectively. Calculate the rate of heat flow through this sphere.

Given

Thermal conductivity of sphere,	k_1 = 65 w/mk
Inner diameter of sphere,	d_1 = 120 mm
Radius,	r_1 = 60 mm = 0.060m
Outer diameter of sphere,	d_2 = 350 mm
Radius	r_2 = 175 mm = 0.175 m
Radius,	$r_3 = r_2$ + thickness of insulation

$$= 0.175 + 0.010$$

$$r_3 \qquad = 0.185 \ m$$

Thermal conductivity of insulation $k_2 = 10 \ w/mk$

Inside temperature, T_a, $= 500°C + 273 = 773 \ k$

Outside temperature $T_b = 50°C + 273 = 323 \ k$

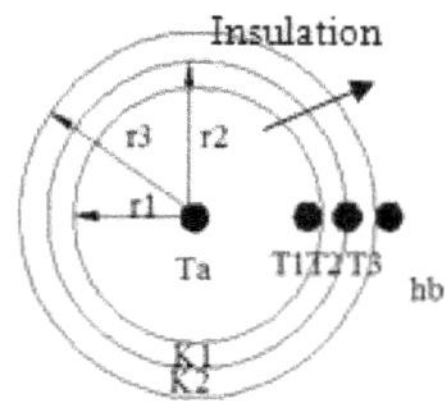

To Find

Heat loss, (Q)

Solution

Step 1

Heat loss, $\quad Q = \dfrac{\Delta T_{overall}}{R}$ [From HMT data book Page.No – 46]

Where, $\Delta T = T_a - T_b$

$$R = \frac{1}{4\pi}\left[\frac{1}{h_a r_1^2} + \frac{1}{k_1}\left[\frac{1}{r_1} - \frac{1}{r_2}\right] + \frac{1}{k_2}\left[\frac{1}{r_2} - \frac{1}{r_3}\right] + \frac{1}{h_b r_3^2}\right]$$

[The terms h_a and h_b are not given, so neglect that terms]

$$\Rightarrow Q = \frac{T_a - T_b}{\dfrac{1}{4\pi}\left[\dfrac{1}{k_1}\left[\dfrac{1}{r_1} - \dfrac{1}{r_2}\right] + \dfrac{1}{k_2}\left[\dfrac{1}{r_2} - \dfrac{1}{r_3}\right]\right]}$$

$$= \frac{773 - 323}{\dfrac{1}{4\pi}\left[\dfrac{1}{65}\left[\dfrac{1}{0.060} - \dfrac{1}{0.175}\right] + \dfrac{1}{10}\left[\dfrac{1}{0.175} - \dfrac{1}{0.185}\right]\right]}$$

$$\Rightarrow Q = 28361 \ w$$

Result

Heat transfer, Q = 28361 w.

1.2. Critical Radius of Insulation

Addition of insulation increases the conductive resistance doesnot reduce the amount of heat transfer rate always. But decrease the convective resistance of the outer surface increases the heat loss up to certain thickness of insulation. The radius of insulation for which the heat transfer is maximum is called critical radius of insulation and corresponding thickness is called critical thickness. If the thickness is further increased, the heat loss will be reduced.

1.2.1. *Critical Radius of Insulation for a Cylinder*

Step 1

Consider a cylinder having thermal conductivity k, Let r_1 and r_o inner and outer radi of insulation.

$$\text{Heat transfer, } Q = \frac{T_i - T_\alpha}{\dfrac{\ln\left(\dfrac{r_0}{r_1}\right)}{2\pi kL}}$$

Considering h be the outside heat transfer co-efficient

$$\therefore Q = \frac{T_i - T_\alpha}{\dfrac{\ln\left(\dfrac{r_0}{r_1}\right)}{2\pi kL} + \dfrac{1}{A_o h}} \qquad\qquad \rightarrow \qquad (1.37)$$

Here $A_a = 2\pi r_0 L$

$$Q = \frac{T_i - T_\alpha}{\dfrac{\ln\left(\dfrac{r_0}{r_1}\right)}{2\pi kL} + \dfrac{1}{2\pi r_o Lh}} \qquad\qquad \rightarrow \qquad (1.38)$$

Step 2

To find the critical radius of insulation, differentiate Q with respect to r_0 and equate it to Zero.

$$\Rightarrow \quad \frac{dQ}{dr_0} = \frac{0 - (T_i - T_\alpha)\left[\dfrac{1}{2\pi kLr_0} - \dfrac{1}{2\pi hLr_0^2}\right]}{\dfrac{1}{2\pi kL}\ln\left(\dfrac{r_0}{r_1}\right) + \dfrac{1}{2\pi kLr_0}}$$

Since $(T_i - T_\alpha) \neq 0$

$$\Rightarrow \frac{1}{2\pi kLr_0} - \frac{1}{2\pi hLr_0^2} = 0$$

$$\Rightarrow r_0 = \frac{k}{h} = r_c \qquad\qquad \rightarrow \qquad (1.39)$$

Table 1.4: Required Formulas for Critical Radius of Insulation

1	Critical radius 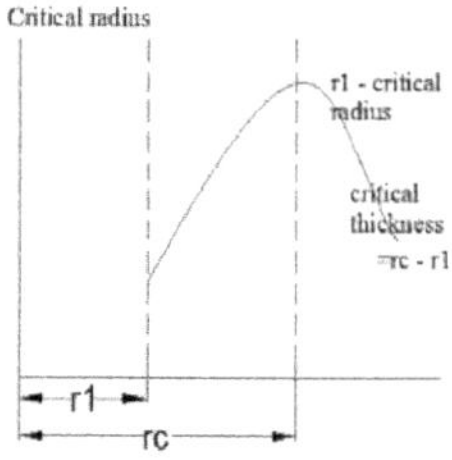	$rc = \dfrac{K}{h}$ $t_c = r_c - r_1$	Pg.No – 48
2	Heat transfer through an insulated wire is given by 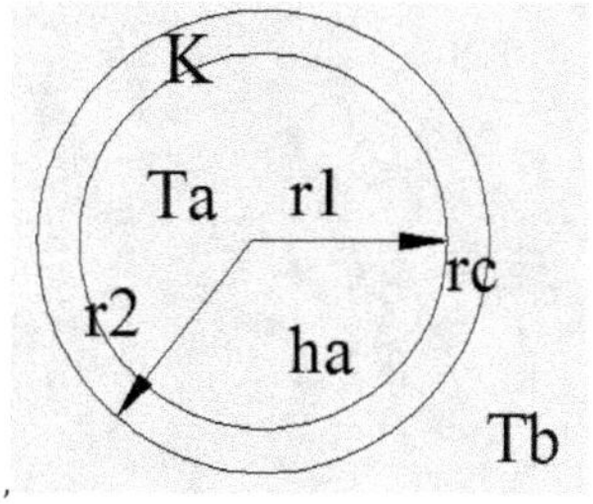	$q = \dfrac{(\Delta T)\,overall}{R}$ $Q_1 = \dfrac{T_a - T_b}{\dfrac{1}{2}\pi L\left[\ln\dfrac{\left(\dfrac{r_2}{r_1}\right)}{K_1} + \dfrac{1}{hbr_2}\right]}$ (or) Without insulation : $Q1 = \dfrac{2\pi L(Ta - Tb)}{\dfrac{1}{har1}}$	Pg.No – 43 to 45
3	Heat transfer through an insulated wire when critical radius is used by	$(r_2\,replaced\,by\,r_c)$ $Q_2 = \dfrac{T_a - T_b}{\dfrac{1}{2}\pi L\left[\ln\dfrac{\left(\dfrac{r_c}{r_1}\right)}{K_1} + \dfrac{1}{hbr_c}\right]}$	Pg.No – 43 to 45
4	Percentage of increase in the heat disscipation ,	$Cr = \dfrac{Q_2 - Q_1}{Q_1} * 100$	

1.2.2. *Solved Problem on Critical Thickness*

1. A wire of 6 mm diameter with 2 mm thick insulation (k = 0.11 w/mk). If the convective heat transfer co-efficient between the insulating surface and air is 25 w/m²k, find the critical thickness of insulation and also find the percentage of change in the heat transfer rate if the critical radius is used.

Given

d_1 = 6 mm

r_1 = 3 mm = 0.003 m

$r_2 = r_1 + 2 = 3 + 2$

= 5 mm = 0.005 m

k = 0.11 w/mk

h_b = 25 w/m²k

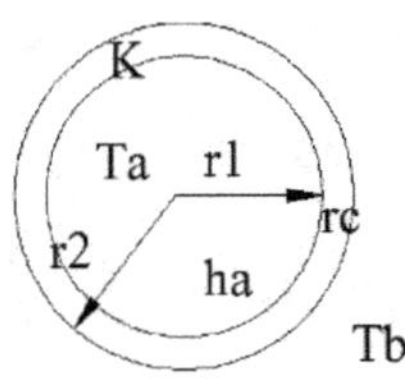

To Find

1. Critical thickness
2. % of change in heat transfer

Solution

Step 1

1. Critical radius, $r_c = \dfrac{k}{h}$

$$r_c = \dfrac{0.11}{25}$$

r_c = 4.4 x 10⁻³ m

Critical thickness, $t_c = r_c - r_1$ = 4.4 x 10⁻³ – 0.003

t_c = 1.4 x 10⁻³ m

Step 2

2. Heat transfer through an insulated wire is given by

[From HMT Data book Pg.No. 43 to 45]

$$Q_1 = \cfrac{T_a - T_b}{\cfrac{1}{2\pi L}\left[\cfrac{ln\left(\cfrac{r_2}{r_1}\right)}{k_1} + \cfrac{1}{h_b r_2}\right]} = \cfrac{2\pi L(T_a - T_b)}{\cfrac{ln\left(\cfrac{0.005}{0.003}\right)}{0.11} + \cfrac{1}{25 \times 0.005}}$$

$$Q_1 = \cfrac{2\pi L(T_a - T_b)}{12.64}$$

Step 3

3. Heat flow through an insulated wire when critical radius is used is given by,

$\therefore [r_2 \rightarrow r_c]$

$$Q_2 = \cfrac{(T_a - T_b)}{\cfrac{1}{2\pi L}\left[\cfrac{ln\left(\cfrac{r_c}{r_1}\right)}{k_1} + \cfrac{1}{h_b r_c}\right]}$$

$$= \cfrac{2\pi L(T_a - T_b)}{\cfrac{in\left(\cfrac{4.4 \times 10^{-3}}{0.003}\right)}{0.11} + \cfrac{1}{25 \times 4.4 \times 10^{-3}}}$$

$$Q_2 = \cfrac{2\pi L(T_a - T_b)}{12.572}$$

Step 4

$\therefore$ Percentage of increase in heat flow by using

Critical radius $= \cfrac{Q_2 - Q_1}{Q_1} \times 100$

$$= \cfrac{\cfrac{1}{12.57} - \cfrac{1}{12.64}}{\cfrac{1}{12.64}} \times 100 = 0.55 \ \%$$

Result

1. Critical thickness, $t_c = 1.4 \times 10^{-3}$ m

2. % of increase in heat transfer by using critical radius = 0.55 %

2. A wire of 7 mm diameter is covered with an insulating material (k= 1 w/mk). The wire temperature ambient temperature are 80°C and 15°C. if the inside convective heat transfer co-efficient is 8.2 w/m²k, find the minimum thickness of insulation and also find the percentage of increase in the heat dissipation.

Given

$d_1 = 7$ mm

$r_1 = 3.5$ mm $= 3.5 \times 10^{-3}$ m

$k = 1$ w/mk

$T_a = 80°C + 273 = 353$ k

$T_b = 15°C + 273 = 288$ k

$h_a = 8.2$ w/m²k

To Find

1. Minimum thickness of insulation
2. % of increase in the heat dissipation

Solution

Step 1

1. Critical radius, $r_c = \dfrac{k}{h}$

$$r_c = \dfrac{1}{8.2}$$

$r_c = 0.1219$ m

Critical thickness, $t_c = r_c - r_1$

$$= 0.1219 - 3.5 \times 10^{-3}$$

$t_c = 0.118$ m

Step 2

2. Heat loss without insulation

$$Q_1 = \dfrac{2\pi L (T_a - T_b)}{1/h_a r_1}$$

$$Q_1 = \dfrac{2\pi L (353 - 288)}{1/8.2 \times 3.5 \times 10^{-3}}$$

$$\dfrac{Q_1}{L} = 11.73 \text{ w/m}$$

Step 3

3. Heat loss with insulation

$$Q_2 = \frac{2\pi L (T_a - T_b)}{\dfrac{\ln\left(\dfrac{r_c}{r_1}\right)}{k_1} + \dfrac{1}{h_a r_c}}$$

$$\frac{Q_2}{L} = \frac{2\pi(353 - 288)}{\dfrac{\ln\left(\dfrac{0.1219}{3.5 \times 10^{-3}}\right)}{1} + \dfrac{1}{8.2 \times 0.1219}}$$

$$\frac{Q_2}{L} = 89.74 \text{ w/m}$$

Step 4

$\therefore$ Percentage of increase in heat dissipation

$$= \frac{Q_2 - Q_1}{Q_1} \times 100$$

$$= \frac{89.74 - 11.72}{11.72} \times 100$$

$$= 665.69\%$$

Result

1. Minimum insulation thickness, $t_c = 0.118$ m

2. % of increase in heat dissipation = 665.69 %

1.3. Heat Conduction with Heat Generation

In many practical cases, there is a heat generation with in the system. Typical Examples are

1. Electric coils

2. Resistance heater

3. Nuclear reactor

4. Combustion of fuel in the fuel bed of boiler furnaces.

In electric coil and resistance heater, heat is generated due to electric current flowing in the fire. In nuclear fuel element, heat is generated by nuclear fission.

1.3.1. *Plane Wall with Internal Heat Generation*

Step 1

Consider a slab of thickness L, thermal conductivity k, as shown in fig 1.3. Consider a small elemental area of thickness d_x.

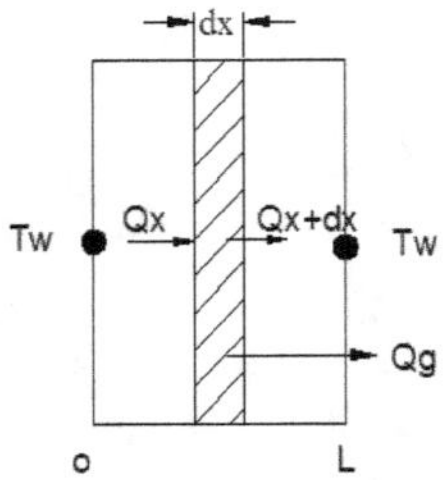

Figure 1.4

Step 2

From Fourier's law of conduction, we know that

Heat transfer at x, $Q_x = -kA\dfrac{dT}{dx}$

Heat conducted out at x + dx

$$Q_{x+dx} = \frac{dT}{dx} - kA\frac{d_T}{d_x}$$

Heat generated within dx

$Q_g = q^{\bullet}Adx$

We knowthat,

$Q_x + Q_g = Q_x + d_x$

$$\Rightarrow -kA\frac{dT}{dx} + q^{*}Adx = -kA\frac{dT}{dx} - kA\frac{d^2T}{dx^2}dx$$

$$\Rightarrow q^{\bullet}Adx = -kA\frac{d^2T}{dx^2}$$

$$\Rightarrow kA\frac{d^2T}{dx^2} + q^{\bullet}Adx = 0$$

$$\therefore \text{ Divided by Ka} \Rightarrow \frac{d^2T}{dx^2} + \frac{q^{\bullet}}{k}d_x = 0$$

Step 3

Integrating above equation

$$\Rightarrow \int \frac{d^2T}{dx^2} + \int \frac{\overset{\bullet}{q}}{k}\,dx = \int 0$$

$$\Rightarrow \frac{dT}{dx} + \frac{\overset{\bullet}{q}}{k}x = C_1$$

Integrating

$$\Rightarrow \int \frac{dT}{dx} + \frac{\overset{\bullet}{q}}{k}\int x = \int C_1$$

$$\Rightarrow T + \frac{\overset{\bullet}{q}}{k}\frac{x^2}{2} = C_1 x + C_2$$

$$\Rightarrow T = -\frac{1}{2}\frac{\overset{\bullet}{q}}{k}x^2 + C_1 x + C_2 \qquad\qquad \to (1.40)$$

Step 4

The temperature on the two faces of the slab (T_w) is the same because it loses the same amount of heat by convection on two sides.

Apply boundary condition; $C_1 = 0$

$$\Rightarrow T = \frac{1}{2}\frac{\overset{\bullet}{q}}{k}\,x_2 + C_2$$

Apply $T = T_w$, $x = L/2$

$$\Rightarrow T_w = -\frac{1}{2}\frac{\overset{\bullet}{q}}{k}\left[\frac{1}{2}\right]^2 + C_2$$

$$C_2 = T_w + \frac{1}{2}\frac{\overset{\bullet}{q}}{k}\left[\frac{L}{2}\right]^2 + C_2$$

$$C_2 = T_w + \frac{\overset{\bullet}{q}L^2}{8k}$$

Substituting C_1 and C_2 value in equation (1.40)

$$T = -\frac{1}{2}\frac{\overset{\bullet}{q}}{k}x^2 + o + T_w + \frac{\overset{\bullet}{q}L^2}{8k}$$

$$\Rightarrow T = T_w + \frac{\overset{\bullet}{q}}{8k}\,(L^2 - 4x^2)$$

The maximum temperature t_{max} (at the centre) is obtained by puting $x = 0$ in equation (2)

$$T_{max} = T_w + \frac{\overset{\bullet}{q}L^2}{8k}$$

Step 5

Heat flow rate

$$Q = \frac{1}{2}\, q^{\bullet} AL$$

Heat transfer by convection

$$Q = hA\,(T_w - T_\infty)$$

$$\Rightarrow Q = \frac{1}{2}\, q^{\bullet} AL = hA(T_w - T_\infty)$$

$$\Rightarrow \frac{1}{2}\, q^{\bullet} AL = hA\,(T_w) - hA\,T_\infty$$

$$\Rightarrow ha\,T_w = hAT_\infty + \frac{1}{2}\, q^{\bullet} AL$$

$$\Rightarrow T_w = T_\infty + \frac{q^{*}L}{2h}$$

Surface or wall temperature

$$T_w = T_\infty + \frac{q^{\bullet}L}{2h} \qquad\qquad\rightarrow\qquad (1.41)$$

1.3.2. *Cylinder with Internal Heat Generation*

Step 1

Consider a cylinder of radius r and thermal conductivity k. Heat is generated (Q_g) in the cylinder due to passage of an electric current.

Step 2

From Fourier's law of conduction, we know that

$$r\frac{d^2 T}{dr^2} + \frac{q^{\bullet} r}{k} = 0$$

Integrating

$$r\int \frac{d^2 T}{dr^2} + \frac{q^{\bullet} r}{k} = \int 0$$

$$\Rightarrow r\frac{dT}{dr} + \frac{q^{\bullet}}{k}\,\frac{r^2}{2} = C_1$$

$$\Rightarrow \frac{dT}{dr} + \frac{q^{\bullet} r}{2k} = \frac{C_1}{r}$$

Integrating

$$\int \frac{dT}{dr} + \int \frac{q^{\bullet}}{2k}\, r = \int \frac{C_1}{r}$$

$$\Rightarrow T + \frac{q^\bullet}{2k}\frac{r^2}{2} = C_1 \ln r + C_2$$

$$\Rightarrow T = -\frac{q^\bullet r^2}{4k} + C_1 \ln r + C_2 \qquad\qquad \rightarrow \quad (1.42)$$

Apply boundary conditions

$C_1 = 0$ [Put $T = T_w$ and $r = r_0$]

$$\Rightarrow \quad T_w = -\frac{q^\bullet r^2}{4k} + C_2$$

$$\Rightarrow \quad C_2 = T_w + \frac{q^\bullet r^2}{4k}$$

Apply C_1 and C_2 value in equation (1.42)

$$T = \frac{-q^\bullet r^2}{4k} + 0 + T_w + \frac{q^\bullet r_o^2}{4k}$$

$$T = T_w + \frac{q^\bullet}{4k}[r_0^2 - r^2]$$

At centre

$r = 0, T = T_{max}$

$$\Rightarrow T_{max} = T_w + \frac{q^\bullet}{4k}[r_0]^2$$

Maximum Temperature, $T_{max} = T_w + \dfrac{q^\bullet r^2}{4k}$

Step 3

We know that,

Heat generated

$$Q = \pi\, r_0^2\, L q^\bullet$$

Heat transfer due to convection

$$Q = hA(T_w - T_\infty)$$

$$Q = h \times 2\pi\, r_0 L\, (T_w - T_\infty)$$

Equating above equating

$$\pi\, r_0^2\, L q^\bullet = h \times 2\,\pi r_0 L\, (T_w - T_\infty)$$

$$r_0 q^\bullet = h \times 2 \times (T_w - T_\infty)$$

$$r_0 q^\bullet = 2h\, T_w - 2h\, T_\infty$$

$$2h T_w = r_0\, q^* + 2h\, T_\infty$$

$$T_w = T_\infty + \frac{r_o q^*}{2h}$$

Surface temperature, $T_w = T_\infty + \dfrac{r_o \dot{q}}{2h}$ $\qquad \rightarrow \qquad$ (1.43)

Similarly

For sphere, temperature at the centre

$$T_c = T_w + \frac{\dot{q} r_0^2}{6k} \qquad \rightarrow \qquad (1.44)$$

<table>
<tr><td colspan="5" align="center">Table 1.5: Internal Heat Generation Formula Wed</td></tr>
<tr><td>S.NO</td><td colspan="2">DESCRIPTION</td><td>FORMULA</td><td>HMT DBP NO</td></tr>
<tr><td>1</td><td colspan="2">Plane wall with heat generation 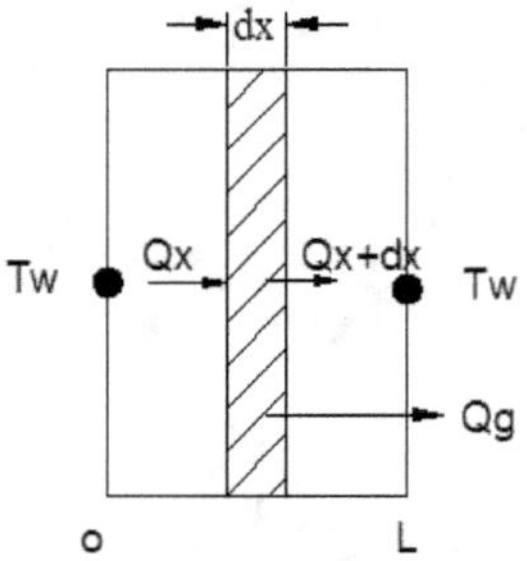</td><td>Surface temperature
$Tw = T\infty + \dfrac{\dot{q}L}{2h}$
Maximum temperature
$T_{max} = Tw + \dfrac{\dot{q}L^2}{8K}$</td><td>Pg.No - 48</td></tr>
<tr><td>2</td><td colspan="2">For cylinder with heat generation:</td><td>1)heat generation
$\dot{a} = \dfrac{Q}{V}$
2)heat transfer
$Q = I^2 R.$
3)Maximum temperature :
$T_{MAX} = Tw + \dfrac{\dot{q}r^2}{4K}$
4)surface temperature
$Tw = T\infty + \dfrac{r\dot{q}}{2h}$
Volume $V = \pi r^2 L$
Radius $r = (m)$</td><td>Pg.no- 48</td></tr>
<tr><td>3</td><td colspan="2">Heat generation for sphere,</td><td>1) Temperature at the center.
$Tc = Tw + \dfrac{\dot{q}r^2}{6K}$
$\dot{q} = \dfrac{Q}{V} \quad , \quad V = \dfrac{4}{3\pi r^3}$</td><td>Pg.No – 49</td></tr>
</table>

1) T∞ - fluid temperature , K

2) Q - heat generation W/m³

3) L - thickness (m)

4) H - heat transfer co-efficient, W/m².K

5) K - thermal conductivity , W/m.k

6) I - current

7) R – Resistance

1.3.3. Solved Problems on Plane Wall with Internal Heat Generation

1. An electric current is passed through a plane wall of thickness 150mm which generates heat at the rate of 50,000 w/m³. The convective heat transfer co-efficient between wall and ambient air is 65 w/m²k. Ambient air temperatures is 28°C and the thermal conductivity of the wall material is 22 w/mk. Calculate:

   1. Surface temperature

   2. Maximum temperature in the wall

Given

Thickness, L = 150mm = 0.150m

Heat generation, $q^{\bullet}$ = 50,000 w/m³

Convective heat transfer coefficient, h = 65 w/m²k

Ambient air temperature, T_{∞} = 28°C + 273 = 301 k

Thermal conductivity, k = 22 w/mk

To Find

1. Surface temperature

2. Maximum temperature in the wall

Solution

Step 1

We know that

Surface temperature

$$T_w = T_{\infty} + \frac{q^{\bullet}}{2h}$$

$$= 301 + \frac{50,000 \times 0.150}{2 \times 65}$$

$$T_w = 358.6 \text{ k}$$

Step 2

Maximum temperature

$$T_{max} = T_w + \frac{q^{\bullet}L^2}{8k}$$

$$= 358.6 + \frac{50,000 \times (0.150)^2}{8 \times 22}$$

$T_{max} = 364.9 \text{ k}$

Result

1. Surface temperature, $T_w = 358.6k$

2. Maximum temperature, $T_{max} = 364.9k$

2. A concrete wall of 1m thick is poured with concrete. The hydration of concrete generates 150 w/m³ heat. If both the surfaces of the wall are maintained at 35°C. Find the maximum temperature in the wall.

Given

Thickness, L = 1m

Heat generation, $q^{\bullet}$ = 150 w/m³

Surface temperature, T_w = 35°C + 273 = 308 k

To Find

Maximum temperature in the wall

Solution

Step 1

Maximum temperature

$$T_{max} = T_w + \frac{q^{\bullet}L^2}{8k}$$

Thermal conductivity of concrete, k = 1.279 w/mk

[From HMT data book pg.no.19]

$$T_{max} = 308 + \frac{150 \times (1)^2}{8 \times 1279 \times 10^{-3}}$$

$T_{max} = 322.6k$

Result

Maximum temperature, $T_{max} = 322.6k$

1.3.4. Solved Problems on Cylinder with Internal Heat Generation

1. A copper wire of 40mm diameter carries 250A and has a resistance of $0.25 \times 10^{-4}\,\Omega$cm/L surface temperature of copper wire is 250°C and the ambient air temperature is 10°C. If the thermal conductivity of the copper wire is 175 w/mk, calculate,

    1. Heat transfer co-efficient between wire surface and ambient air.
    2. Maximum temperature in the wire.

Given

Diameter, d = 40mm = 0.04m

Radius, r = 20mm = 0.020m

Current, I = 250A

Resistance, $R = 0.25 \times 10^{-4}\,\Omega$ cm/length

Surface temperature. $T_w = 200°C + 273 = 523\,k$

Ambient Air temperature, $T_\infty = 10°C + 273 = 283k$

Thermal conductivity, k = 175 w/mk

To Find

1. Heat transfer co-efficient, h
2. Maximum temperature, T_{max}

Solution

Step 1

Heat transfer, $Q = I^2R$

$$= (250)^2 \times (0.25 \times 10^{-4})$$

Q $= 1.562$ w/cm $= 1.56 \times 10^2$ w/m

Q $= 156$ w/m

Step 2

We know that,

Heat generated, $q° = \dfrac{Q}{V}$ $[v = \pi r^2 L]$

$$q^{\bullet} = \frac{156}{\pi \times (0.020)^2 \times 1}$$

$q^{\bullet} = 124140$ w/m³

Step 3

We know that,

Maximum temperature.

$$T_{max} = T_w + \frac{\dot{q}\,r^2}{4k}$$

$$= 523 + \frac{124140 \times (0.020)^2}{4 \times 175} = 523.07 \text{ k}$$

$$T_{max} = 523.07 \text{k}$$

Step 4

We know that

Surface temperature, $T_w = T_\infty + \dfrac{r\dot{q}}{2h}$

$$523 = 283 + \frac{0.020 \times 124140}{2 \times h}$$

$$h = 5.17 \text{ w/m}^2\text{k}$$

Result

1. Heat transfer co-efficient, $h = 5.17$ w/m²k
2. Maximum temperature, $T_{max} = 523.07$k.

1.3.5. Solved problems on Sphere with Internal heat Generation

1. A sphere of 100mm diameter having thermal conductivity of 0.18 w/mk. The outer surface temperature is 8ₒC. And 250 w/m² of energy is released due to heat source. Calculate

    1. Heat generated
    2. Temperature at the centre of the sphere

Given

Diameter of sphere, d = 100mm

r = 50mm = 0.050 m

Thermal conductivity, k = 0.18 w/mk

Surface temperature, T_w = 8ₒC + 273 = 281 k

Energy released, Q = 250 w/m²

To Find

1. Heat generated, $\overset{\bullet}{q}$
2. Temperature at the centre of the sphere

Solution

Step 1

Heat generated, $\overset{\bullet}{q} = \dfrac{Q}{V}$

$\Rightarrow \quad \dfrac{\overset{\bullet}{q}}{A} = \dfrac{Q/A}{V}$ $\qquad\qquad\qquad$ $[\because Q/A = 250 \ w/m^2]$

$\Rightarrow \quad \dfrac{\overset{\bullet}{q}}{A} = \dfrac{250}{4/3 \ \pi r^3}$ $\qquad\qquad\qquad$ $[V = 4/3 \ \pi \ r^3]$

$\Rightarrow \dfrac{\overset{\bullet}{q}}{4\pi r^2} = \dfrac{250}{4/3 \ \pi r^3}$

$\Rightarrow q^o = \dfrac{250 \times 4 \times \pi \times (0.050)^2}{4/3 \times \pi \times (0.050)^3}$

$\Rightarrow q^o = 15{,}000 \ w/m^3$

Step 2

Temperature at the centre of the sphere

$T_c \quad = T_w + \dfrac{\overset{\bullet}{q} r^2}{6k}$

$\quad = 281 + \dfrac{15{,}000 \times (0.050)^2}{6 \times 0.18}$

$T_c \quad = 315.7 \ k$

Result

1. Heat generated, $\overset{\bullet}{q} = 15{,}000 \ w/m^2$
2. Centre temperature, $T_c = 315.7k$

1.4. Transient Heat Conduction (Or) Unsteady State Conduction

If the system temperature of a body does not vary with time, it is said to be in a steady state. But if there is an abrupt change in its surface temperature, it attains a steady state after some period. During this period the temperature varies with time and the body is said to be in an un steady or transient state.

Transient heat conduction occurs in cooling of I_c engines, automobiles, boiler tubes, heating and cooling of metal billets, rocket nozzles, electric irons etc.

Transient heat conduction can be divided in to periodic heat flow and non periodic heat flow.

(i) Periodic Heat Flow

In periodic heat flow, the temperature varies on a regular basis.

Example: Cylinder of an Icengine

Surface of earth during period of 24 hours

(ii) Non Periodic Heat Flow

In non periodic heat flow, the temperature at any point within the system varies non – linearly with time.

Example: Heating of an in got in a furnace cooling of bars.

1.4.1. Solved Problem on a Unsteady State Conduction

1. The temperature distribution across a large concrete slab [k = 1.2 w/m°c, α = 1.77 x 10^{-3} m^2/h] 500mm thick heated from one side as measured by thermo couples approximates to the relations T = 60 – 50x + 12x^2 + 20x^3 - 15x^4 where 'T' is in °C and x is in meters. Considering an area of 5m^2 compute.

    (i)   The heat entering and leaving the slabs in unit times.

    (ii) The heat energy stored in unit time

    (iii) The rate of temperature change at both side of slabs

    (iv) The point where the rate of heating or cooling is maximum

Given

K = 1.2 w/m°C

$$\alpha = 1.77 \times 10^{-3} \ m^2/hr \ = \ \frac{1.77 \times 10^{-3}}{3600}$$

α = = 4.91 x 10^{-7} m^2/sec

$x = 500mm = 0.5m$

$A = 5m^2$

$T = 60 - 50x + 12x^2 + 20x^3 - 15x^4$

Solution

Step 1

$T = 60 - 50x + 12x^2 + 20x^3 - 15x^4$

Diff. w.r.t x,

$$\frac{dT}{dx} = -50 + 24x + 60x^2 - 60x^3$$

$$\frac{d^2T}{dx^2} = 24 + 120x - 180x^2$$

$$\frac{d^3T}{dx^3} = 120 - 360x$$

Step 2

(i) Heat Entering the Slab

$$Q_{in} = -kA \left(\frac{dT}{dx} \right)_{x=0}$$

$$= -1.20 \times 5 \times [-50 + 0]$$

$Q_{in} = 300w$

Heat leaving the slab

$$Q_{out} = -kA \left(\frac{dT}{dx} \right)_{x=0.5}$$

$$= 1.20 \times 5 \times [-50 + (24 \times 0.5) + (60 \times 0.5^2) - (60 \times 0.5^3)]$$

$Q_{out} = 183w$

Step 3

(ii) Heat Energy Stored in Unit Time

$$Q = Q_{in} - Q_{out}$$

$$= 300 - 183$$

$$Q = 117 \, w$$

Step 4

(iii) Rate of Temperature Change at Both Sides of the Slab

$$\frac{dT}{dt} = \alpha \frac{d^2T}{dx^2}$$

$$= 4.91 \times 10^{-7} [24 + 120x - 180x^2]$$

For Entering

$$\left(\frac{dT}{dt}\right)_{x=0} = 4.91 \times 10^{-7} [24 + 0]$$

$$= 1.17 \times 10^{-5}\,°C/sec$$

For Leaving

$$\left(\frac{dT}{dt}\right)_{x=0.5} = 4.91 \times 10^{-7} [24 + (120 \times 0.5) - (180 \times 0.5^2)]$$

$$= 1.91 \times 10^{-5}\,°c/sec$$

Step 5

iv) The Point where the Rate of Heating or Cooling is Maximum

$$\frac{d}{dx}\left[\alpha \frac{d^2T}{dx^2}\right] = 0$$

$$\alpha \frac{d^3T}{dx^3} = 0$$

$$\frac{d^3T}{dx^3} = 0$$

$$120 - 360x = 0$$

$$12 = 360x$$

$$x = 0.333m$$

2. At a certain instant of time, the temperature distribution in a long cylindrical tube is, T = 800 + 1000r − 5000r² where, T is in °C and r in m. The inner and outer radii of the tube are respectively 30cm and 50cm. The tube material has a thermal conductivity of 58 w/m.k and a thermal diffisivity of 0.004 m²/hr. Determine (i) the rate of heat flow at inside and outside surfaceper length (ii) Rate of heat storage per unit length (iii) Rate of change of temperature at inner and outer surfaces.

Given

In cylindrical tube,

$T = 800 + 1000r - 5000r^2$

Inner radius, $r_1 = 30cm = 30 \times 10^{-2}m$

Outer radius, $r_2 = 50cm = 50 \times 10^{-2}m$

Thermal conductivity, $k = 58$ w/mk

Thermal diffusivity, $\alpha = 0.004$ m^2/hr

$$= \frac{0.004}{3600} = 1.11 \times 10^{-6} \text{ m}^2/\text{s}$$

To Find

(i) Rate of heat flow at inside and outside surfaces per unit length

(ii) Rate of heat storage per unit length

(iii) Rate of change of temperature at inner and outer surfaces.

Solution

Step 1

(i) Rate of Heat Flow at Inside Surface Per Unit Length

$$Q_{in} = -k\, A_i \left(\frac{dT}{dr} \right)_{r_i = 0.3} \qquad\qquad [A_i = 2\pi r L]$$

$$Q_{in} = -58 \times 2\pi \times (0.3) \times 1 \times \left(\frac{dT}{dr} \right)_{ri=0.3}$$

$$T = 800 + 1000r - 5000r^2$$

$$\frac{dT}{dr} = 1000 - 10000r$$

$$\frac{d^2T}{dr^2} = -10{,}000$$

$$Q_{in} = -109.33\,[1000 - 10000\,(0.3)]$$

$$= -109.33\,[-2000]$$

$$Q_{in} = 21.86 \times 10^4 \text{w}$$

Rate of Heat Flow at Outside Surface Per Unit Length

$$Q_{out} = -k\, Ao \left(\frac{dT}{dr} \right)_{r_0 = 0.5}$$

$$= -58 \times 2\pi \times (0.5) \times [1000 - 10000\,(0.5)]$$

$$= -58 \times 3.14 \times [-4000]$$

$$Q_{out} = 72.84 \times 10^4\, w$$

Step 2

(ii) Rate of Heat Storage Per Unit Length

$$\therefore Q_{stored} = Q_{in} - Q_{out} = (21.86 - 72.84) \times 104$$

$$Q_{stored} = -50.98 \times 10^4\, w$$

Step 3

(iii) Rate of Change of Temperature at Inner Surfaces at $r_i = 0.3m$

$$\frac{d^2T}{dr^2} + \frac{1}{r}\frac{dT}{dr} = \frac{1}{\alpha} \cdot \frac{dT}{dt}$$

$$\left(\frac{dT}{dt} \right)_{r_i = 0.3} = \alpha.\left[\frac{d^2T}{dr^2} + \frac{1}{r}\frac{dT}{dr} \right] = 1.11 \times 10^{-6}\left[-10000 + \frac{1}{0.3}(1000 - (10000 \times 0.3)) \right]$$

$$\left(\frac{dT}{dt} \right)_{r_i = 0.3} = 0.01851\,°c/S$$

Rate of change of temperature at outer surfaces

$$\left(\frac{dT}{dt} \right)_{r_i = 0.5} = \alpha.\left[\frac{d^2T}{dr^2} + \frac{1}{r}\frac{dT}{dr} \right]$$

$$= \left(\frac{dT}{dt} \right)_{r_0 = 0.5} 1.11 \times 10^{-6}\left[-10000 + \frac{1}{0.5}(1000 - (10000 \times 0.5)) \right]$$

$$\left(\frac{dT}{dt} \right)_{r_i = 0.5} = -0.02\,°c/S$$

1.4.2. Biot Number

The ratio of internal conduction resistance to the surface convection resistance is known as Biot number.

$$\text{Biot Number} = \frac{\text{Intenral conduction resis}\,tan\,ce}{\text{Surface convection resis}\,tan\,ce}$$

$$Bi = \frac{hL_c}{K}$$

Where, k – Thermal conductivity, w/mk

H – Heat transfer co-efficient, w/m²k

T_c – characteristics length or significant length

$$L_c = \frac{\text{Volume}}{\text{Surface Area}} = \frac{V}{A}$$

(i) For slab, L_c = L/2, L – Thickness of the slab

(ii) For cylinder, L_c = R/ 2, R – Radius of cylinder

(iii) For sphere, L_c = R/3, R – Radius of Sphere

(iv) For cube, L_c = L/6, L – Thickness of the cube

1.4.3. *Lumped Heat Analysis [Negligible Internal Resistance]*

The process in which the internal resistance is assumed as negligible in comparison with its surface resistance is known as Newtonian heating or cooling process.

In a Newtonian heating or cooling process the temperature is considered to be uniform at a given time. Such an analysis is called Lumped parameter analysis.

Step 1

Let us consider a solid whose initial temperature is T_0 and it is placed suddenly in ambient air or any liquial at a constant temperature T_∞. The transient response of the body can be determined by relating its rate of change of internal energy with convective exchange at the surface.

Qconv = E

$$= -hA\,(T - T_\infty) = pcV\frac{dT}{dt}$$

T = T_0 at t = 0

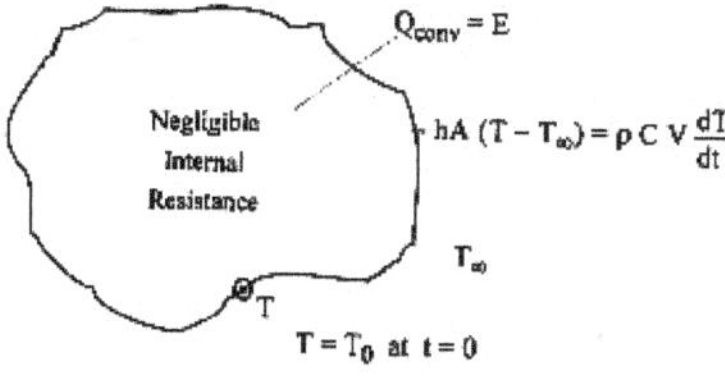

Lumped Heat Capacity System

Step 2

$$\text{Convective heat loss from the body} \quad = \quad \text{Rate of change of internal energy}$$

$$-hA(T - T_\infty) = \rho \times c_p \times v\frac{dT}{dt}$$

$$\frac{dT}{T - T_\infty} = \frac{-hA}{\rho C_p \, v} \times dt$$

Integrating

$$\int \frac{dT}{T - T_\infty} = \frac{-hA}{\rho C_p \, v} \times \int dt$$

$$\Rightarrow \ln [T - T_\infty] = \frac{-hA}{\rho C_p \, v} t + C_1 \text{ ----- (1)} \qquad\qquad \rightarrow \qquad (1.45)$$

$$C_1 = \ln [T_0 - T_\infty]$$

Substituting c_1 value in equation (1.45)

$$\Rightarrow \ln [T - T_\infty] = \frac{-hA}{\rho C_p \, v} t + \ln [T_0 - T_\infty]$$

$$\Rightarrow \ln [T - T_\infty] - \ln [T_0 - T_\infty] = \frac{-hA}{e C_p \, v} t$$

$$\Rightarrow \ln \left[\frac{T - T_\infty}{T_0 - T_\infty} \right] = \frac{-hA}{\rho \, Cp \, v} t$$

$$\Rightarrow \frac{T - T_\infty}{T_0 - T_\infty} = e^{\left[\frac{-hAt}{\rho Cp \, v} \right]} \qquad\qquad \rightarrow \qquad (1.46)$$

Where,

T_0 = Initial temperature of the solid k

T = Intermediate temperature of the solid, k

T_∞ = Surface temperature (or) Final temperature of the solid k

h = Heat transfer co-efficient, w/m^2k

A = Surface area of body, m^2

P = Density of the body, kg/m^3

V = Volume of the body, m^3

C_p = Specific heat of the body, J/kgk

T = time, s.

Note

In lumped parameter system, Biot number value is less than 0.1

$$B_i < 0.1 \Rightarrow \frac{hL_c}{k} < 0.1$$

Table 1.6: Transient Heat Conduction (or) Unsteady State Conduction			
S.NO	DESCRIPTION	FORMULA	HMT DBP NO
1	Lumped parameter system : (Bi<0.1)	$Bi = \dfrac{hLc}{k}$ $L_c = \dfrac{V}{A} = \dfrac{volume}{surface\ area}$	Pg.No – 58
	For slab : Characteristic (Lc)	$Lc = \dfrac{V}{A} = \dfrac{AxL}{2A}$, $Lc = \dfrac{L}{2}$ $L - Thickness\ of\ the\ slab$	
	For cylinder : Characteristic (Lc)	$Lc = \dfrac{V}{A} = \dfrac{\pi R^2 L}{2\pi RL}$, $Lc = \dfrac{R}{2}$ $R - Radius\ of\ cylinder$	
	For sphere Characteristic (Lc)	$Lc = \dfrac{V}{A} = \dfrac{\frac{4}{3}\pi R^3}{4\pi R^2}$, $Lc = \dfrac{R}{3}$ $R - Radius\ of\ sphere$	
	For cube Characteristic (Lc)	$Lc = \dfrac{V}{A} = \dfrac{L^3}{6L^2} = \dfrac{L}{6}$ $L - thickness\ of\ the\ cube$	
2	For lumped parameter sustem ,	$\dfrac{T - T\infty}{To - T\infty} = e\left[-\dfrac{hA}{ecpV} * t\right]$	Pg. No 58

T - Intermediate temp in K

T - Temp of solid at the time in K

To - Initial temp in K

T∞ - Final temp in K

e – Density, Kg/m^3

Cp - Specific heat j/kg.k

t - time, sec

1.4.4. *Solved Problem on a Lumped Heat Analysis*

1. A copper plate 2mm thick is heated upto 400°C and quenched in to water at 30°C. Find the time required for the plat to reach the temperature of 50°C. Heat transfer co-efficient is 100 w/m²k, density of copper is 8800 kg/m³. Specific heat of copper = 0.36 kJ/kgk
   Plate dimensions = 30 x 30 cm

Given

Thickness of plate, L = 2mm = 0.002m

Initial temperature, T_0 = 400°C + 273 = 673k

Final temperature, T_∞= 30°C + 273 = 303k

Intermediate temperature, T = 50°C + 273 = 323k

Heat transfer Co-efficient, h = 100 w/m²k

Density, p = 8800 kg/m³

Specific heatCp = 360 J/kg k

Plate dimensions = 30 x 30 cm

To Find

Time required for the plate to reach 50°C

Solution

Step 1

Thermal conductivity of the copper, k = 386 w/mk.

[From HMT Data book page. No. 2 and 3]

For Slab

Characteristic length, L_c $\qquad = \dfrac{L}{2} = \dfrac{0.002}{2}$

$\qquad L_c \qquad = 0.001m$

Step 2

We know that,

Biot number, $B_i \qquad = \dfrac{hLc}{k} = \dfrac{100 \times 0.001}{386}$

$\qquad B_i \qquad = 2.59 \times 10^{-4} < 0.1$

Biot number value is less than 0.1. So, this is lumped heat analysis type problem.

Step 3

For lumped parameter system

$$\frac{T - T_\infty}{T_0 - T_\infty} = e^{\left[\frac{-hA}{Cp \times v \times \rho} \times t\right]}$$

[From HMT D.B. Pg.No. 58]

Characteristics length, $L_c = \dfrac{V}{A}$

$$(1) => \frac{T - T_\infty}{T_0 - T_\infty} = e^{\left[\frac{-h}{Cp \times L_c \times \rho} \times t\right]}$$

$$=> \frac{323 - 303}{673 - 3.3} = e^{\left[\frac{-100}{360 \times 0.001 \times 8800} \times t\right]}$$

$$=> \ln(0.0545) = \left[\frac{-100}{360 \times 0.001 \times 8800} \times t\right]$$

$$=> t = 92.43 \text{ s}$$

Result

Time required for the plate to reach 50°C is 92.43 S

2. A 12cm diameter long bar initially at a uniform temperature of 40°C is placed in a medium at 650°C with a convective co-efficient of 22 w/m²k. Determine the time required for the centre to reach 255°C. For the material of the bar, k = 20 w/mk, Density = 580 kg/m³. Specific heat = 1050 J/kgk.

Given

Diameter of bar, D = 12cm = 0.12m

Radius of bar, R = 6 cm = 0.06 m

Initial temperature, T_0 = 40°C + 273 = 313 k

Final temperature, T_∞ = 650°C + 273 = 923k

Intermediate temperature, T = 255°C + 273 = 528 k

Heat transfer co-efficient, h = 22 w/m²k

Thermal conductivity, k = 20 w/mk

Density, ρ = 580 kg/m³

Specific heat, C_p = 1050 J/kgk

To Find

Time required (t).

Solution

Step 1

For Cylinder

$$\text{Characteristic length, } L_c = \frac{R}{2} = \frac{0.06}{2}$$

$$L_c = 0.03\,m$$

We know that,

$$\text{Biot number, } B_i = \frac{hL_c}{k}$$

$$= \frac{22 \times 0.03}{20}$$

$$B_i = 0.033 < 0.1$$

Biot number value is less than 0.1, so this is lumped heat analysis type problem.

Step 2

For lumped parameter system,

$$\frac{T - T_\infty}{T_o - T_\infty} = e^{\left[\frac{-hA}{Cp \times V \times \rho} \times t\right]}$$

[From HMT Databookpg.No. 58]

$$\text{Characteristic length, } L_c = \frac{T - T_\infty}{T_o - T_\infty} = e^{\left[\frac{-h}{Cp \times L_c \times \rho} \times t\right]}$$

$$= \ln\left[\frac{528 - 923}{313 - 923}\right] = e^{\left[\frac{-22 \times t}{1050 \times 0.03 \times 580} \times t\right]}$$

$$t = 360.8\,S$$

Result

Time required for the cylinder to reach 255°C is 360.8 S

3. Alloy steel ball of 12mm diameter heated to 800°C is quenched in a bath at 100°C. The material properties of the ball are k = 205 KJ/m.hr.k, P = 7860 kg/m³, C_p = 0.45 kJ/kg k, h = 150 kJ/hr. m².k.

   Determine (i) Temperature of ball after 10 seconds and (ii) Time for ball to cool to 400°C.

Given

Diameter of the ball, D = 12mm = 0.012 m

Radius of the ball, R = 0.006 m

Initial temperature, T_0 = 800°C + 273 = 1073k

Final temperature, T_∞= 100°C + 273 = 373k

Thermal conductivity, k = 205KJ/m.hr.k

$$= \frac{205 \times 1000}{3600\,smk}\,J$$

$$= 56.94\ w/mk \quad [\because J/S - w]$$

Density, ρ = 7860 kg/m³

Specific heat, C_p= 0.45 KJ/kgk

$$= 450\ J/kgk$$

Heat transfer Co-efficient, h $\quad$ = 150 KJ/hr. m².k

$$= \frac{150 \times 1000J}{3600\,Sm^2k}$$

$$= 41.66\ w/m^2k$$

To Find

(i)   Temperature of ball after 10 sec

(ii) Time for ball to cool to 400°C

Solution

Step 1

Case (i) Temperature of ball after 10 sec

For sphere

Characteristic length, L_c $\qquad = \dfrac{R}{3}$

$$= \frac{0.006}{3}$$

$$L_c \quad = 0.002m$$

We know that, Biot number, $B_i = \dfrac{hLc}{k}$

$$= \frac{41.667 \times 0.002}{56.94}$$

$$B_i = 1.46 \times 10^{-3} < 0.1$$

Biot number value is less than 0.1. So this is lumped heat analysis type problem.

Step 2

For lumped parameter system,

$$\frac{T - T_\infty}{T_o - T_\infty} = e^{\left[\frac{-hA}{Cp \times V \times \rho} \times t\right]}$$

[From HMT Data Book pg.no.58]

Characteristic length = $L_c = \dfrac{V}{A}$

$$\Rightarrow \frac{T - T_\infty}{T_o - T_\infty} = e^{\left[\frac{-h}{Cp \times L_c \times \rho} \times t\right]}$$

$$\Rightarrow \frac{T - 373}{1073 - 373} = e^{\left[\frac{-41.667}{450 \times 0.002 \times 7860} \times 1.0\right]}$$

$$\Rightarrow T = 1032.95k$$

Step 3

Case (ii) Time for ball to cool to 400°C

$\therefore$ T = 400°C + 273 = 673k

$$\Rightarrow \frac{T - T_\infty}{T_o - T_\infty} = e^{\left[\frac{-h}{Cp \times L_c \times \rho} \times t\right]} \Rightarrow \frac{673 - 373}{1073 - 373} = \left[\frac{-41.667}{450 \times 0.002 \times 7860} \times t\right]$$

$$\Rightarrow \ln\left[\frac{673 - 373}{1073 - 373}\right] = \frac{-41.667}{450 \times 0.002 \times 7860} \Rightarrow t = 143.849 \text{ S}$$

Result

(i) Temperature of ball after 10 sec T = 1032.95k

(ii) Time for ball to cool to 400°C, t = 143.8498 S

4. An aluminium cube 6 cm on a side is originally at a temperature of 500°C. It is suddenly immersed in a liquid at 10°C for which h is 120 w/m²l. Estimate the time required for the cube to reach a temperature of 250°C. For aluminium ρ = 2700 kg/m³, C_p = 900 J/kg k, K = 204 w/mk.

Given

Thickness of cube, L = 6 cm = 0.06 m

Initial temperature, T_0 = 500°C + 273 = 773k

Final temperature, $T_\infty = 10\,^\circ C + 273 = 283k$

Intermediate temperature, $T = 250\,^\circ C + 273 = 523k$

Heat transfer Co efficient, $h = 120\ w/m^2 k$

Density, $\rho = 2700\ kg/m^3$

Specific heat, $C_p = 900\ J/kg\ k$

Thermal conductivity, $k = 204\ w/mk$

To Find

Time required for the cube at reach $250\,^\circ C$

Solution

Step 1

For cube

Characteristic length, $L_c \quad = \dfrac{L}{6}$

$$Lc \quad = 0.01m$$

We know that

Biot number, $B_i \quad = \dfrac{hLc}{k} = \dfrac{120 \times 0.01}{204}$

$B_i \quad = 5.88 \times 10^{-3} < 0.1$

Biot number value is less than 0.1, So this is lumped heat analysis type problem.

Step 2

For lumped parameter system

$$\frac{T-T_\infty}{T_0-T_\infty} = e^{\left[\frac{-hA}{Cp\, x\, V\, x\, \rho} x\, t\right]}$$

[From HMT Databookpg.No. 58]

Characteristic length, $L_c = {V}\big/{A}$

$$=> \frac{T-T_\infty}{T_0-T_\infty} = e^{\left[\frac{-h}{Cp\, x\, L_c\, x\, \rho} x\, t\right]}$$

$$=> \frac{523-283}{773-283} = e^{\left[\frac{-120}{900 \times 0.01 \times 2700} x\, t\right]}$$

$$\Rightarrow \ln (0.489) = \dfrac{-120}{900 \times 0.01 \times 2700}\, xt$$

$$\Rightarrow t = 144.86\ S$$

Step 3

Result

Time required for the cube to reach 250°C is 144.86S

1.4.5. Heat Flow in Semi Infinite Solids

A solid which extends itself infinitely in all directions of space is known as infinite solid. If an infinite solid is split in the middle by a plane, each half is known as semi infinite solid. In a semi infinite solid, at any instant of time, there is always a point where the effect of heating (or cooling) at one of its boundaries is not felt at all. At this point the temperature remains unchanged.

Step 1

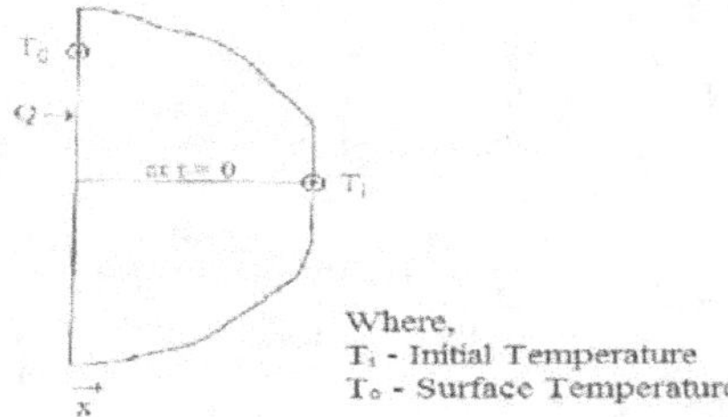

Consider a semi-infinite body and it extends to infinity in the +ve x direction. The entire body is initially at uniform temperature T_i including the surface at x = 0.

The surface temperature at x = 0

Suddenly raised to T_0.

Step 2

The governing equation is $\dfrac{d^2T}{dx^2} = \dfrac{1}{\alpha}\dfrac{dT}{dt}$

The boundary conditions are

1. $T(x, 0) = T_i$
2. $T(0, t) = T_0$ for $t > 0$
3. $T(\alpha, t) = T_i$ for $t > 0$

The analytical solution for this case is given by

$$\frac{T_x - T_0}{T_i - T_0} = \text{erf}\left[\frac{x}{2\sqrt{\alpha t}}\right] \qquad \rightarrow \qquad (1.47)$$

Where,

erf indicates "Error function of" and the definition of error function is generally available in mathematical texts. Usually tabulation of error values are available in data books.

α = Thermal diffusivity, m²/s

t = time, S

X = Distance, m

T_i = Initial temperature, k

T_0 = Surface (or) Final temperature, k

T_x = Intermediate temperature, k

Note

1. In semi infinite solid, heat transfer co-efficient or biot number value is ∞.

i.e., h →∞

or

B_i→∞

<table>
<tr><td colspan="4" align="center">Table 1.7: Heat Flow in Semi–Infinite Solids</td></tr>
<tr><td>S.NO</td><td>DESCRIPTION</td><td>FORMULA</td><td>HMT DB P.NO</td></tr>
<tr><td>1</td><td>Bioth number (Bi = ∞)
There fore (heat transfer coefficient his not given) h = ∞</td><td>$Bi = \dfrac{hLc}{K}$</td><td>Pg.No – 58</td></tr>
<tr><td>2</td><td>For semi infinite solid ,

(values of error function
Z table Pg.No 60)</td><td>$\dfrac{Tx - To}{Ti - To} = \text{erf}\left[\dfrac{x}{2\sqrt{\alpha t}}\right]$

$\dfrac{Tx - To}{Ti - To} = \text{erf}[Z]$

$there\ fore\ \left(Z = \dfrac{x}{2\sqrt{\alpha t}}\right)$</td><td>Pg.No – 58</td></tr>
<tr><td>3</td><td>i)   Instantaneous heat flow ,

ii)   Total heat energy</td><td>$q_x = \dfrac{K[To - Ti]}{\sqrt{\alpha \pi t}} e\left[\dfrac{-x^2}{4\alpha t}\right]$

$q_t = 2K[To - Ti]\left(\dfrac{\sqrt{t}}{\pi \alpha}\right)$</td><td>Pg.No – 59</td></tr>
<tr><td>4</td><td>Heat flux</td><td>$q_o = \dfrac{K[To - Ti]}{\sqrt{\alpha \pi t}}$</td><td>Pg.No – 59</td></tr>
</table>

Tx -   Temp at a distance x from surface at time   t

To -   surface temperature

Ti -   initial temp of solid

α -   Thermal difflisivity of solid m^2/S => K/ecp -(m^2/S)

T -   time

K -   thermal conductivity in w/m.k

1.4.6. Solved Problems – Semi Infinite Solids

1. A large concrete high way initially at a temperature of 70°C and stream water is directed on the high way so that the surface temperature is suddenly lowered to 40°C. Determine the time required to reach 55°C at a depth of 4cm from the surface.

Given

Initial temperature,	T_i	= 70°C + 273 = 343 k
Final or Surface temperature,	T_0	= 40°C + 273 = 313 k
Intermediate temperature,	T_x	= 55°C + 273 = 328 k
Depth, x = 4cm		= 0.04m

To Find

Time (t) required to reach 55°C

Solution

Step 1

Properties of concrete are

[From HMT Data book pg.No. 19]

Thermal conductivity, k = 1.2790 w/mk

Thermal diffusivity, $\alpha = \dfrac{k}{\rho c_p}$

$\therefore$ Density, ρ = 2300 kg/m³

Specific heat, C_p = 1130 J/kgk

$$\alpha = \frac{1.2790}{2300 \times 1130} = 0.49 \times 10^{-6} \text{ m}^2/\text{s}$$

Step 2

In this problem heat transfer Co-efficient h is not given, So that it as ∞.

i.e., h = ∞

Biot number $B_i = \dfrac{hL_c}{k} = \infty$

B_i value is ∞, this is semi infinitesolid type problem.

Step 3

For semi infinite solid,

$$\frac{T_x - T_0}{T_i - T_0} = \text{erf}\left[\frac{x}{2\sqrt{\alpha t}}\right]$$

[From HMT Databook pg. No. 59]

$$\Rightarrow \frac{T_x - T_0}{T_i - T_0} = \text{erf}(z) \quad \text{Where, } z = \frac{x}{2\sqrt{\alpha t}}$$

$$\Rightarrow \frac{328 - 313}{343 - 313} = \text{erf}(z)$$

$$0.5 = \text{erf}(z)$$

$\Rightarrow \text{erf}(z) = 0.5$

erf (z) = 0.5, corresponding z is 0.48   [HMT.b.Pg.No.60]

We know that,

$$Z = \frac{x}{2\sqrt{\alpha t}}$$

$$\Rightarrow 0.48 = \frac{0.04}{2\sqrt{0.49 \times 10^{-6} \times t}}$$

$$\Rightarrow (0.48)^2 = \frac{(0.04)^2}{(2)^2 \times 0.49 \times 10^{-6} \times t}$$

$\Rightarrow t = 3535.8S$

Result

Time required to reach 55°C is 3535.8 S.

2. A very thick was initially at a temperature of 25°C and wall temperature is suddenly raised to 700°C and remains constant there after calculate the following

1. Temperature in plane at a depth of 300mm from the surface after 7 hours.

2. Instantaneous heat flow rate at a depth of 300mm and on surface after 7 hours.

3. Total heat energy after 7 hours.

Taks k = 0.75 w/mk, α = 0.002 m²/hr.

Given

Initial temperature, T_i = 25°C + 273 = 298 k

Final temperature, T_0 = 700°C + 273 = 973 k

Depth, x = 300mm = 0.300 m

Time, t = 7hr = 25,200 S

Thermal diffusivity, α = 0.002 m²/hr = 5.5 x 10⁻⁷ m²/s

Thermal conductivity, k = 0.75 w/mk

To Find

1. Intermediate temperature, T_x
2. Instantaneous heatflow, q_x
3. Total heat energy, q_T

Solution

Step 1

In this problem heat transfer co-efficient h is not given, so take it as ∞.

(i.e.) h→∞

We know that, $B_i = \dfrac{hLc}{k}$, h →∞

$\quad$ => $B_i = \infty$

B_i value is ∞. So, this is semi-infinite solid type problem.

Step 2

1. Forsemi Infinite Solid

$$\frac{T_x - T_0}{T_i - T_0} = erf\left[\frac{x}{2\sqrt{xt}}\right]$$

$$\Rightarrow z = \frac{x}{2\sqrt{\alpha t}}$$

$$= \frac{0.3}{2\sqrt{5.55 \times 10^{-7} \times 25,200}}$$

$Z = 1.27$

$Z = 1.27$, Corresponding erf(z) is 0.92751

$\Rightarrow erf(z) = 0.92751$

[From HMT Data book pg.no.60]

$$\Rightarrow \frac{T_x - T_0}{T_i - T_0} = 0.92751$$

$$\Rightarrow \frac{T_x - 973}{298 - 973} = 0.92751$$

$$\Rightarrow T_x = 346.9 \, k$$

Step 3

2. Instantaneous Heat Flow

$$q_x = \frac{k[T_0 - T_i]}{\sqrt{\alpha \pi t}} e^{\left[\frac{-x^2}{4\alpha t}\right]}$$

[From HMT DatabookPg.No. 48)

$$\Rightarrow q_x = \frac{0.75(973 - 298)}{\sqrt{\pi \times 5.55 \times 10^{-7} \times 25,200}} \times e^{\left[\frac{-(0.3)^2}{4 \times 5.55 \times 10^{-7} \times 25,200}\right]}$$

$$\Rightarrow q_x = 483.36 \, w/m^2$$

Step 4

3. Total Heat Energy

$$q_T = 2k[T_0 - T_i]\sqrt{\frac{t}{\pi\alpha}} = 2 \times 0.75 (973 - 298) \times \sqrt{\frac{25,200}{\pi \times 5.55 \times 10^{-7}}}$$

$$q_T = 121.72 \times 10^6 \, J/m^2$$

Result

1. Temperature at a depth of 300mm $T_x = 346.9$ k
2. Instateneous heat flow, $q_x = 483.36$ w/m^2
3. Total heat energy, $q_T = 121.72 \times 10^6$ J/m^2

3. A semi infinits slab of aluminium is exposed to a constant heat flux at the surface of 0.25 Mw/m^2. Initial temperature of the slab is 25°C. Calculate the surface temperature after 10 minutes and also find the temperature at a distance of 30 cm from the surface after 10 minutes.

Given

Heat flux, $q_0 = 0.25$ Mw/m^2

$q_0 = 0.25 \times 10^6$ w/m^2

Initial temperature, $T_i = 25°C + 273 = 298$ k

Distance, $x = 30$ cm $= 0.30$ m

Time, $t = 10$ minutes $= 600$ s

To Find

1. Surface temperature (T_0) after 10 minutes.
2. Temperature (T_x) at a distance of 30 cm from the surface.

Solution

Step 1

$$\text{Heat flux, } q_0 = \frac{k[T_0 - T_i]}{\sqrt{\alpha \pi t}}$$

[From HMT Data Book page No. 59]

Properties of Aluminium

Thermal diffusivity, $\alpha = 84.18 \times 10^{-6}$ m^2/s

[From HMT Data Book page No. 59]

Thermal conductivity, $k = 204.2$ w/mk

$$\Rightarrow 0.25 \times 10^6 = \frac{204.2\,(T_0 - 298)}{\sqrt{\pi \times 84.18 \times 10^{-6} \times 600}}$$

$T_0 = 785.68$ k

Step 2

For semi infinite solid,

$$\frac{T_x - T_0}{T_i - T_0} = \text{crf}\left[\frac{x}{2\sqrt{\alpha t}}\right]$$

[From HMT Data book page no. 59]

$$=> \frac{T_x - T_0}{T_i - T_0} = \text{crf}(z)$$

Where

$$Z = \frac{x}{2\sqrt{\alpha t}} = \frac{0.30}{2\sqrt{84.18 \times 10^{-6} \times 600}}$$

Z = 0.667

Z = 0.667, corresponding crf (z) is 0.65663

[From HMT Data book pg.no. 60]

erf(z) = 0.65663

$$\frac{T_x - 785.68}{298 - 785.68} = 0.65663$$

=>T_x = 465.45k

Result

1. Surface temperature, T_0 = 785.68k

2. Temperature at a distance of 30cm, T_x = 465.45 k

1.4.7. *Transient Heat flow in an infinite plate*

A solid which extends itself infinity in all directions of space is known as infinite solid.

Step 1

Consider as infinite flat plate of uniform thickness 2L as shown in fig, 1.4 which is initially at a uniform temperature of T_i. It is suddenly exposed to a large mass of fluid having a temperature T_∞. This temperature is assumed to be constant through out the process of cooling or heating the plate is extended to infinity in the y and z directions.

The heat transfer co-efficient between the surface of the plate and the fluid on both sides is assumed to be constant. The centre of the plate is selected as the region,

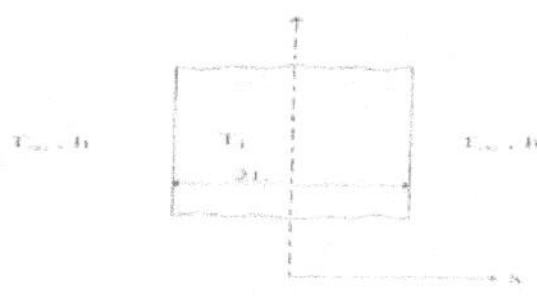

Infinite Plate

Step 1

The governing differential equation is

$$\frac{d^2 T}{dx^2} = \frac{1}{\alpha} \frac{dT}{dx}$$

The boundary conditions are

1. At $t = 0$, $T_0 = T_i$

2. At $x = 0$, $\dfrac{dT}{dx} = 0$

3. At $x = \pm L$, $KA \dfrac{dT}{dx} = hA\,[T_0 - T_\infty]$

The solution of the above differential equation with these boundary condition is given by.

$$\frac{T_0 - T_\infty}{T_i - T_\infty} = f\left[\frac{x}{L},\ \frac{hL}{k},\ \frac{\alpha t}{L^2}\right]$$

From this equation, we know that conduction resistance is not negligible. The temperature history becomes a function of biot number $\left[\dfrac{hLc}{k}\right]$, Fourier number $\left[\dfrac{\alpha t}{L^2}\right]$ and the dimensionless parameter $\left[\dfrac{x}{L}\right]$ which indicates the location of point with in the plate where temperature is to be obtained. The dimensionless parameter $\left[\dfrac{x}{L}\right]$ is replaced by $\left[\dfrac{r}{R}\right]$ in case of cylinders and spheres.

Heisler Charts

Heisler has prepared charts for graphical solutions of the consteady state conduction problems. These charts have been constructed in non-dimensional parameters. The charts are suitable for problems with a finite surface and internal resistance. For such case biot number lies between 0 and 100. These heiler charts were further extended and improved by grober. The heislerand grober charts are used to solve the problems of sudden immersion of plate, cylinder or sphere in to a fluid.

Note

In infinite solids, biot number value is in between 0.1 to 100. i.e., $0.1 < B_i < 100$.

<table>
<tr><td colspan="4" align="center">Table 1.8: Transient Heat Flow in Infinite Plates</td></tr>
<tr><td>S.NO</td><td>DESCRIPTION</td><td>FORMULA</td><td>HNT DB P.NO</td></tr>
<tr><td>1</td><td>Bi = biot number values in between 0.1 and 100 Ie ., 0.1<Bi < 100</td><td>$Bi = \dfrac{hLc}{K}$</td><td>Pg.No – 58</td></tr>
<tr><td>2</td><td>Fourier number, (x – axis) (Heister charts)

$Y_{axis} = \left(\dfrac{T_o - T_\infty}{T_i - T_\infty}\right)$</td><td>$\dfrac{at}{Lc^2}$
$\dfrac{hLc}{K}$</td><td>Pg.No – 66</td></tr>
<tr><td>3</td><td>X axis
Curve

Y axis</td><td>$Bi = \dfrac{hLc}{K}$
$\dfrac{x}{Lc}$
$\dfrac{Tx - T\infty}{To - T\infty}$</td><td>Pg.No - 67</td></tr>
</table>

1.4.8. *Solved Problems on Infinite Solids*

1. A slab of aluminium 10 cm thick is originally at a temperature of 500°C. It is suddenly immersed in a liquid at 100°C resulting in a heat transfer co-efficient of 1200 w/m²k. Determine the temperature at the centre line and the surface 1 minute after the immersion. Also calculate the total thermal energy removed per unit area of the slab during this period. The properties of aluminium for the given conditions are

   $\alpha = 8.4 \times 10^{-5}$ m²/s $\qquad$ k = 215 w/mk

   P = 2700 kg/m³ $\qquad$ c = 0.9 kJ/kgk

Given

Thickness, L = 10cm = 0.10 m

Initial temperature, T_i = 500°C + 273 = 773k

Final temperature, T_∞ = 100°C + 273 = 373k

Heat transfer co-efficient, h = 1200 w/m²k

Properties of aluminium,

$P = 2700 \text{ kg/m}^3$

$\alpha = 8.4 \times 10^{-5} \text{ m}^2/\text{s}$

$K = 215 \text{ w/mk}$

$C_p = 0.9 \text{ kJ/kgk} = 0.9 \times 10^3 \text{ J/kgk}$

To Find

1. Temperature at centreline after 1 minutes
2. Temperature at the surface
3. Total thermal energy removed per unit area

Solution

Step 1

For slab,

$$\text{Characteristic length, } L_c = \frac{L}{2} = \frac{0.10}{2}$$

$$L_c = 0.05 \text{m}$$

$$\text{Biot number, } B_i = \frac{hLc}{L} = \frac{1200 \times 0.05}{215}$$

$\Rightarrow B_i = 0.279$

Biot number value is in between 0.1 and 100

i.e., $0.1 < B_i < 100$, so, this is infinite solid type problem.

Step 2

Case (i)

Mid plane temperature for infinite plate.

Heisler chart [From HMT Data book pg.no.66]

$$\text{X axis} \rightarrow \text{Fourier number} = \frac{\alpha t}{L_c^2} = \frac{8.4 \times 10^{-5} \times 60}{(0.05)^2}$$

$[\therefore t = 1 \text{ minute} = 60\text{s}]$

$\text{X axis} \rightarrow \text{Fourier number} = 2.016$

$$\text{Curve} \rightarrow \quad = \frac{hLc}{k}$$

$$= \frac{1200 \times 0.05}{215} = 0.279$$

$$\text{Curve} \rightarrow = \frac{hLc}{k} = 0.279$$

X axis value is 2.016, curve value is 0.279 corresponding y axis value is 0.64.

$$\frac{T_o - T_\infty}{T_i - T_\infty}$$

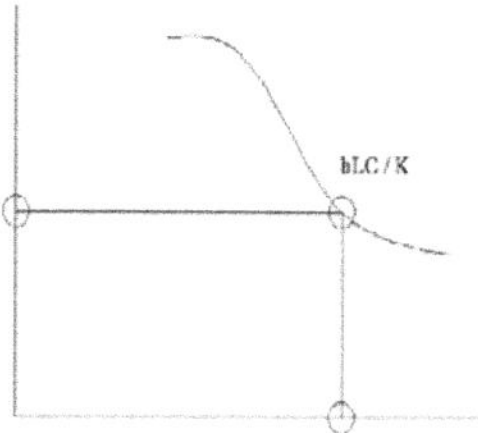

$$\frac{\alpha t}{L_c^{\,2}} = 2.061$$

$$\text{Y axis} = \frac{T_0 - T_\infty}{T_i - T_\infty} = 0.64$$

$$=> \frac{T_0 - 373}{773 - 373} = 0.64$$

$T_0 = 629k$

Centre line temperature, $T_0 = 629k$

Step 3

Case (ii)

Temperature at the surface $X = L_c$

$=> x = 0.05m$

[From HMT Databook page no. 67]

- Haislerchart

$$\text{X axis} \rightarrow \text{Biot number, } B_i = \frac{hLc}{k} = 0.279$$

$$\text{Curve} \rightarrow = \frac{x}{Lc} = \frac{Lc}{Lc} = \frac{0.05}{0.05} = 1$$

X axis value is 0.279, curve value is 1, from that we can find corresponding y axis value is 0.88.

$$\frac{T_x - T_\infty}{T_0 - T_\infty} = 0.88$$

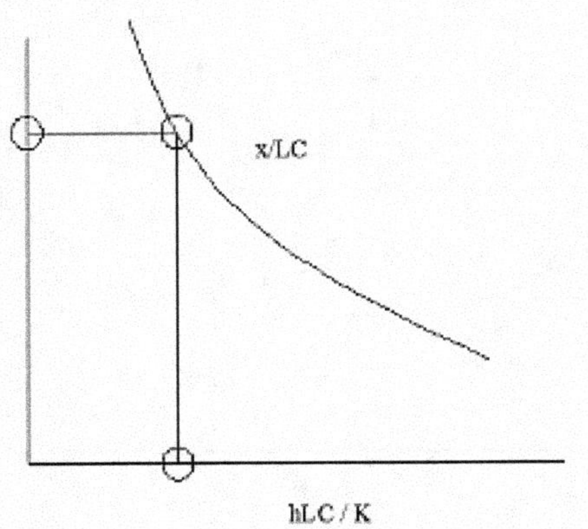

Y axis = $\dfrac{T_x - T_\infty}{T_0 - T_\infty}$ = 0.88

=> $\dfrac{T_x - 373}{629 - 373} = 0.88$

=> $T_x = 598.28k$

Temperature at a surface, T_x= 598.28k

Step 4

Case (iii)

Total thermal energy removed (or) total heat energy removed.

[Refer HMT Data book pg.no.67-Heisler chart]

X axis → Fourier number $= \dfrac{h^2 \alpha t}{k^2}$

$= \dfrac{(1200)^2 \times 8.4 \times 10^{-5} \times 60}{(215)^2}$

X axis = 0.157

Curve = $\dfrac{hLc}{k} = \dfrac{1200 \times 0.05}{215} = 0.279$

X axis value is 0.157, curve value is 0.279 from that, we can find corresponding axis value is 0.34.

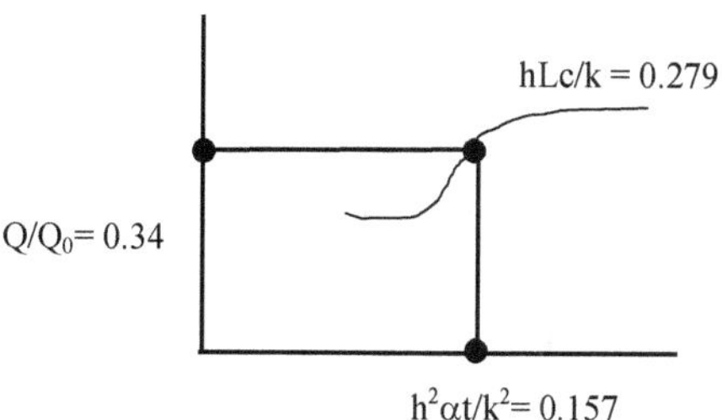

Y axis $\dfrac{Q}{Q_0}$ = 0.34

We know that,

$Q_0 = \rho c_p \, L \, [T_i - T_\infty]$

[Refer HMT Databook pg.no. 64]

$\quad = 2700 \times 0.9 \times 10^3 \times 0.10 \,[773 - 373]$

$Q_0 \quad = 97.2 \times 10^6 \text{ J/m}^2$

$=> Q \qquad = 0.34 \times Q_0$

$=> Q \qquad = 0.34 \times 97.2 \times 10^6$

$Q \quad = 33.04 \times 10^6 \text{ J/m}^2$

Result

1. Temperature at the centre line, $T_0 = 629k$
2. Temperature at the surface, $T_x = 598.28k$
3. Total thermal energy removed per unit are
    $Q = 33.04 \times 10^6 \text{ J/m}^2$

2. A long cylinder 12cm diameter and initially at 20°C is placed in to furnace at 820°C with h = 140 w/m²k. Calculate the time required for the axis temperature to reach 800°C. Also calculate the corresponding temperature at a radius of 5.4 cm at the time physical properties of steel are k = 21 w/mk, α = 6.11 x 10⁻⁶ m²/s.

Given

Diameter of cylinder D = 12 cm = 0.12m

Radius R = 6 cm = 0.06m

T_i = 20°C + 273 = 293 k

T_∞ = 820°c + 273 = 1093k

Heat transfer co-efficient, h = 140 w/m²k

Axis temperature

Or $\Bigg\}$ T_o = 800°C + 273 = 1073k

Centre line temperature

Intermediate radius, r = 5.4 cm = 0.054m

Thermal diffusivity, α = 6.11 x 10⁻⁶ m²/s

Thermal conductivity, k = 21 w/mk.

To Find

1. Time (t) required for the axis temperature to reach 800°C
2. Corresponding temperature T^r at radius of 5.4 cm

Solution

Step 1

For Cylinder

Characteristic length $L_c = \dfrac{R}{2} = \dfrac{0.06}{2}$

L_c = 0.03m

We know that,

Biot number, $B_i = \dfrac{hLc}{k}$

$$= \dfrac{140 \times 0.03}{21}$$

=> $B_i = 0.2$

0.1 < Bi < 100, So, this is infinite solid type problem.

Step 2

Case (i)

Axis temperature T_o = 800°C + 273 = 1073k

[Refer HMT Data book pg.no. 69]

Curve = $\dfrac{hR}{k}$

$$= \frac{140 \times 0.06}{21} = 0.4$$

$$\text{Y axis} = \frac{T_o - T_\infty}{T_i - T_\infty} = \frac{1073 - 1093}{293 - 1093}$$

Y axis = 0.025

Curve value is 0.4, y axis 0.025. From that, we can find corresponding x axis value is 5.

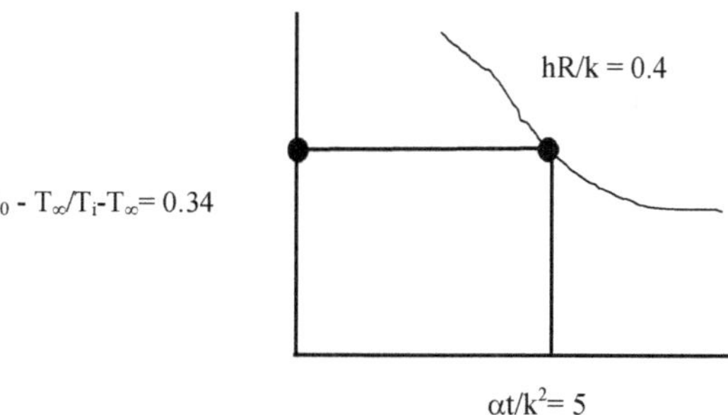

$$\Rightarrow \text{X axis} = \frac{\alpha t}{R^2}$$

$$\Rightarrow t = \frac{5\,R^2}{\alpha}$$

$$= \frac{5 \times (0.06)^2}{(6.11 \times 10^{-6})}$$

t = 2945.9 S

Step 3

Case (ii)

Intermediate radius, r = 5.4 cm = 0.054m

[Refer HMT Databook pg. No. 70]

$$\text{Curve} = \frac{r}{R} = \frac{0.054}{0.06} = 0.9$$

$$\text{X axis} = \frac{hR}{k} = \frac{140 \times 0.06}{21} = 0.4$$

Curve value is 0.9, x axis value is 0.4, From that, we can find corresponding y axis value is 0.84.

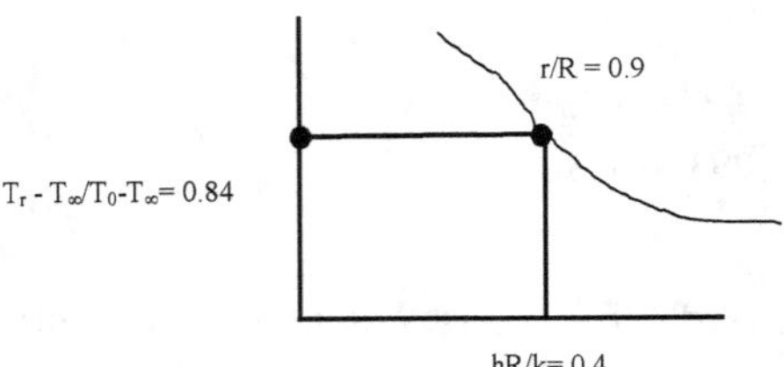

$$\text{Y axis} = \frac{T_r - T_\infty}{T_0 - T_\infty} = 0.84$$

$$=> \frac{T_r - 1093}{1073 - 1093} = 0.84$$

$T_r = 1076.2 \text{ k}$

Result

1. Time required for the axis temperature to reach 800°C is 2945.9S

2. Temperature (T_r) at a radius of 5.4 cm is 1076.2k

3. A 10 cm diameter apple approximately spherical in shape is taken from 20°C environment and placed in a refrigerator where temperature is 5°C and average heat transfer co-efficient is 6w/m²k calculate the temperature at the centre of the apple after a period of 1 hour. The physical properties of apple are density = 998 kg/m³. Specific heat = 4180 J/kg k, Thermal conductivity = 0.6 w/mk.

Given

Diameter of sphere, D = 10cm = 0.10m

Radius of sphere, R = 5cm = 0.05m

Initial temperature, T_i = 20°C + 273 = 293k

Final temperature, T_∞ = 5°C + 273 = 278 k

Time, t = 1 hour = 3600 S

Density, p = 998 kg/m³

Heat transfer co-efficient, h = 6 w/m²k

Specific heat, C_p = 4180 J/kg k

Thermal conductivity, k = 0.6 w/mk

$$\text{Thermal diffusivity, } \alpha = \frac{k}{\rho c_p} = \frac{0.6}{998 \times 4180}$$

$\alpha = 1.43 \times 10^{-7} \text{ m}^2/\text{s}$

To Find

Centre of line temperature (T_o)

Solution

Step 1

For sphere

Characteristic length, $L_c = \dfrac{R}{3} = \dfrac{0.05}{3}$

$L_c = 0.016\,m$

We know that

Biot number, $B_i = \dfrac{hLc}{k} = \dfrac{6 \times 0.016}{0.6}$

$\qquad => B_i = 0.16$

$0.1 < B_i < 100$, so, this is infinite solid type problem.

Step 2

[Refer HMT Data book pg.no. 72]

$\text{X axis} = \dfrac{\alpha t}{R^2} = \dfrac{1.43 \times 10^{-7} \times 3600}{(0.05)^2}$

$\text{X axis} = 0.20$

$\text{Curve} = \dfrac{hR}{k} = \dfrac{6 \times 0.05}{0.6} = 0.5$

X axis value is 0.20m curve value is 0.5 from that, we can find corresponding

Y axis value is 0.86

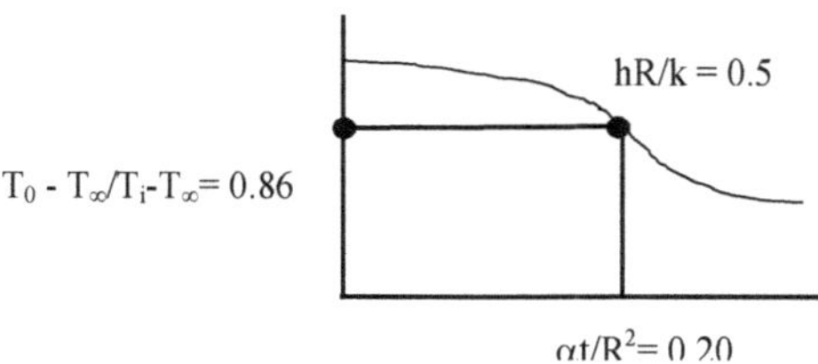

$=> \text{Y axis} = \dfrac{T_o - T_\infty}{T_i - T_\infty} = 0.86$

$=> \dfrac{T_o - 278}{293 - 278} = 0.86$

$=> T_0 = 290.9\,k$

Result

Centre line temperature, $T_0 = 290.9\,k$

Table 1.9: Fins (or) Extended Surfaces Formula Used

S.NO	DESCRIPTION	FORMULA	HMT DBP .NO
1	Infinitely long fin (or) long fin a) Temperature distribution b) Heat transfer	$\dfrac{(T-T\infty)}{T_b-T\infty} = e^{-mx}$ $therefore, m = \dfrac{\sqrt{hp}}{KA}$ $Q = (Tb - T\infty)\sqrt{hpKA}$	Pg.No – 50
2	Short fin (end is insulated) [L/d < 30] a) Temperature distribution	$\dfrac{T - T\infty}{Tb - T\infty} = -\cos\dfrac{hm[L - x]}{\cos h\,[mL]}$	Pg.No – 50
	b) Heat transformed	$Q = (hpKA)^{\frac{1}{2}}(Tb - T\infty)\tan h\,(mL)$	Pg.No – 50
	c) Fin efficiency	$n = \dfrac{\tanh(mL)}{mL}$	Pg.No – 50
	d) Effectiveness	$e = \dfrac{\tanh(mL)}{\dfrac{\sqrt{hA}}{KP}}$	Pg.No – 50

Where,

Tb	–	Base temp in K
T∞	-	surrounding temp in K
T	–	Intermediate temp K
X	-	Distance in m
H	-	Heat transfer coefficient $W/m^2.K$
P	-	Perimeter in m
K	-	Thermal conductivity W/m.K
A	-	Area in m^2

1.4.9. Solved Problems on Extended Surface

1. One end of a very long aluminium rod is connected to a wall at 140°C while the other and extended in to a room. Whose air temperature 15°C. The rod is 3mm in diameter and the heat transfer co-efficient between the rod surface and environment is 300 w/m^2k estimate total heat dissipated by the rod taking thermal conductivity as 150 w/mk.

Given

Long find because long rod

$T_b = 140°C + 273 = 413k$

T_∞ = 15°C + 273 = 288k

D = 3mm = 3 x 10⁻³ m

H = 300 w/m²k

K = 150 w/mk

To Find

Heat dissipate d Q

Solution

Step 1

[From HMT Databook page no. 50]

$$Q = (T_b - T_\infty) \sqrt{hpkA} \ (w)$$

Perimeter, $p = \pi d = \pi \times 3 \times 10^{-3} = 9.424 \times 10^{-3}$m

$$\text{Area A} = \frac{\pi}{4}d^2 = \frac{\pi}{4} \times (3 \times 10^{-3})^2 = 7.06 \times 10^{-6} \text{ m}^2$$

$$Q = (413 - 288)\sqrt{300 \times 9.424 \times 10^{-3} \times 150 \times 7.06 \times 10^{-6}}$$

Q = 6.843w

Result

Heat dissipated Q = 6.843w

2. An aluminium rod (k = 204 w/mk) 2cm in diameter and 20cm long protrudes from a wall which is maintained at 300°C. The end of the rod isinsulated and the surface of the rod is exposed to air at 30°C. The heat transfer co-efficient between the rod surface and air is 10 w/m²k. Calculate the heat lost by the rod and the temperature of the rod at a distance of 10cm from the wall.

Given

K = 204 w/mk

Diameter, d = 2 cm = 0.02m

Length L = 20cm = 0.20m

Base temperature, T_b = 300°C + 273 = 573k

Surrounding temperature, T_∞ = 30°C + 273 = 303k

Hpat transfer co-efficient, h =10 w/m^{2}k

To Find

1. Heat lost by the rod, Q
2. Temperature of the rod at a distance of 10cm from the wall.

Solution

Step 1

$$\frac{L}{d} = \frac{0.20}{0.02} = 10 < 30$$

So, this is short fin

We know that,

Heat transferred [Short fin, end insulated]

$Q = [hpkA]^{\frac{1}{2}} (T_b - T_\infty) \tan h \, (mL)$

[From HMT Databook page. No. 50]

Where,

Area – A = $\dfrac{\pi}{4}d2 = \dfrac{\pi}{4}(0.020)^2$ = 3.14 x 10^{-4}m^2

Perimeter – p = πd = π x 0.02 = 0.0628m

$$m = \sqrt{\frac{hp}{kA}}$$

$$= \sqrt{\frac{10 \times 0.0628}{204 \times 3.14 \times 10^{-4}}}$$

m = 3.13m^{-1}

Q = (10 x 0.0628 x 204 x 3.14 x 10^{-4})$^{\frac{1}{2}}$ x (573 – 303) x $\tan h$ (3.13 x 0.20)

Q = 30.07 w

Step 2

We know that,

Temperature distribution [Short fin, and insulated]

$$\frac{T - T_\infty}{T_b - T_\infty} = \frac{\text{Cosh}\left[m(L-x)\right]}{\cosh\,(mL)}$$

[From HMT Databook page no.50]

Put x = 10cm = 0.10m

$$\Rightarrow \frac{T-303}{573-303} = \frac{\cosh\left[3.13(0.20-0.10)\right]}{\cosh\,(3.13\times0.20)}$$

$$\Rightarrow \frac{T-303}{573-303} = 0.8727$$

$$\Rightarrow T = 538.63k$$

Result

1. Heat lost by the rod, Q = 30.07
2. Temperature of the rod at a distance of 10cm from the wall = 538.63k.

3. A circumferential rectangular profile fin on a pipe of 50mm outer diameter is 3mm thick and 20mm long. Thermal conductivity is 45 w/mk. Convect is co-efficient is 100 w/m²k. Base temperature is 120ºC and surrounding air temperature is 35ºC Determine.

   1. Heat flow rate per f in
   2. Fin efficiency
   3. Fin effectiveness

Given

Diameter, d = 50m =0.050m

Thickness, t = 3mm = 0.003m

Length, L = 20mm = 0.020m

Thermal conductivity, k = 45 w/mk

Convective co-efficient, h = 100 w/m²k

Base temperature, T_b = 120ºC + 273 = 393k

Surrounding temperature, T_∞ = 35ºC + 273 = 308k

To Find

1. Heat flow rate per fin, Q
2. Fin efficiency, η
3. Fin effectiveness, E

Solution

Step 1

$$\frac{L}{d} = \frac{0.020}{0.050} = 0.4 < 30$$

It is treated as short fin, Assume end is insulated

Heat transferred [Short fin, end insulated]

$Q = (hpkA)^{1/2} (T_b - T_\infty) \tan h \, (mL)$

(From HMT Databook page no. 50)

Where,

P = perimeter = $\pi d = \pi \times 0.050 = 0.157\,m$, A – Area = $\dfrac{\pi}{4}d^2 = \dfrac{\pi}{4} \times (0.050)^2 = 1.96 \times 10^{-3}\,m^2$,

Step 2

$$M = \sqrt{\frac{hP}{KA}} = \sqrt{\frac{100 \times 0.157}{45 \times 1.96 \times 10^{-3}}}$$

$M = 13.3\ m^{-1}$

$Q = [100 \times 0.157 \times 45 \times 1.96 \times 10^{-3}]^{1/2} \times (393 - 308) \times \tan h \, (13.3 \times 0.0200)$

$Q = 25.9\ W$

Step 3

Fin efficiency, $\eta = \dfrac{\tan h \, (mL)}{mL}$

$$\eta_{fin} = \frac{\tan h \, (13.3 \times 0.020)}{13.3 \times 0.020} = 97.7\%$$

Step 4

Fin effectiveness

$$E = \frac{\tan h\, mL}{\sqrt{\dfrac{hA}{KP}}} = \frac{\tan h \, (13.3 \times 0.020)}{\sqrt{\dfrac{100 \times 1.96 \times 10^{-3}}{45 \times 0.157}}} = 1.56$$

Result

1. Heat flow Q = 25.9w

2. Fin efficiency, η = 97.7%

3. Fin effectiveness, E = 1.56

4. A cylinder 1m long and 5cm in diameter is placed in an atmosphere at 45°C. It is provided with 10 longitudinal straight fins of material having k = 120 w/mk. The height of 0.76 mm thick fin is 1.27 cm from the cylinder surface. The heat transfer co-efficient between cylinder and atmosphere air is 17 w/m²k. Calculate the rate of heat transfer and the temperature at the end of fins if surface temperature of cylinder is 150°C.

Given

Length of engine cylinder, L_{cy} = 1m

Diameter of the cylinder, d = 5cm = 0.05m

Atmosphere temperature, T_∞ = 45°C + 273

$T_\infty = 318k$

Number of fins = 10

Thermal conductivity of f in, $k = 120$ w/mk

Thickness of f in, $t = 0.76$ mm

$= 0.76 \times 10^{-3}$ m

Length (height) of the fin, $Lf = 1.27$ cm $= 1.27 \times 10^{-2}$ m

Heat transfer co-efficient, $h = 17$ w/m²k

Cylinder surface temperature

Or $\left.\phantom{\begin{array}{c}a\\b\\c\end{array}}\right\}$ T_b $= 150°C + 273$ $= 423$ K

Base temperature

To Find

1. Rate of heat transfer, Q
2. Temperature at the end of the f in

Solution

Step 1

Length of fin is 1.27 cm, so, this is short f in. assuming that the f in end is insulated.

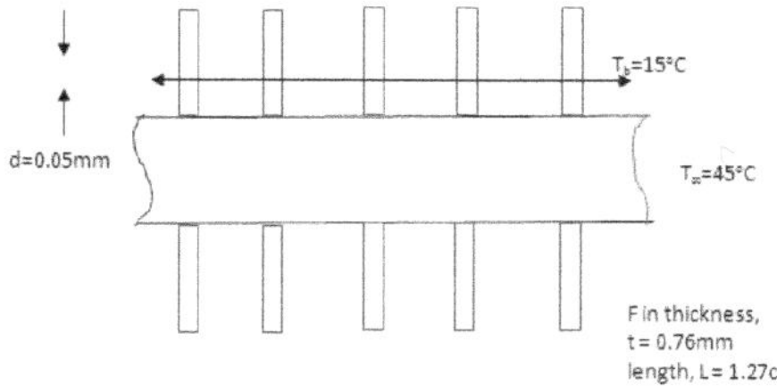

We know that,

Heat transferred,

$Q_1 = (hpKA)^{\frac{1}{2}} (T_b - T_\infty) \tan h(mL_f)\ldots (1)$

Where, P – Perimeter = 2 x length of the cylinder

$= 2 \times 1$

P $= 2$m

A – Area $=$ Length of cylinder x thickness

$= 1 \times 0.76 \times 10^{-3}$

A $= 0.76 \times 10^{-3}$ m²

M $= \sqrt{\dfrac{hP}{KA}}$

$= \sqrt{\dfrac{17 \times 2}{120 \times 0.76 \times 10^{-3}}}$

M $= 19.30\,m^{-1}$

Step 2

Q $= (hpKA)^{\frac{1}{2}} (T_b - T_\infty)\, \tan h\, (mLf)$

$= (17 \times 2 \times 120 \times 0.76 \times 10^{-3})^{\frac{1}{2}} \times (423 - 318) \times \tan h\, (19.30 \times 1.27 \times 10^{-2})$

$= 1.76 \times 10^5 \times 0.24$

Q_1 $= 44.3\ w$

Heat transferred per fin = 44.3w

Heat transferred (Q_1) for 10 fin = 10 x 44.3 = 443 w

Step 3

Heat transfer from unfinned surface due to convections

Q_2 $= hA\Delta T = h\,(\pi d\, Lcy - 10 \times t \times L_f) \times (T_b - T_\infty)$

[$\because$ Area of unfinned surface = Area of cylinder – Area of fin]

$= 17 \times [(\pi \times 0.05 \times 1) - (10 \times 0.76 \times 10^{-3} \times 1.27 \times 10^{-2})] \times (423 - 318)$

Q_2 $= 280.21w$

Step 4

So, Total heat transfer Q

Q $= Q_1 + Q_2$

$= 443 + 280.21$

Q $= 723.21w$

Step 5

We know that

Temperature distribution [Short fin, end insulated]

$$\frac{T - T_\infty}{T_b - T_\infty} = \frac{\cosh\left[m(L_f - x)\right]}{\cosh(mL_f)}$$

[From HMT Databook page no. 50]

We need temperature at the end of fin

So, put x = L

$$\Rightarrow \frac{T - T_\infty}{T_b - T_\infty} = \frac{\cosh\left[m(L-L)\right]}{\cosh(mL_f)}$$

$$\Rightarrow \frac{T - 318}{423 - 318} = \frac{1}{\cosh(19.30 \times 1.27 \times 10^{-2})}$$

$$\frac{T - 318}{423 - 318} = \frac{1}{1.030}$$

$$T = 419.94k$$

Result

1. Heat transfer, $Q = 723.21w$
2. Temperature at end of the fin $T = 419.94k$

5. A circumferential aluminium fin thermal conductivity 200 w/mk of rectangular profile C1.5cm wide and 1mm thick) are fitted on to a 2.5 cm diameter tube. The fin base temperature is 170°C and ambient temperature is 25°C. Estimate heat loss per fin the heat transfer co-efficient (h) may be taken as 130 w/m²k

Given

Circumferential f in, rectangular f in is long and wide, so, heat loss is calculated by using fin efficiency curves.

[From HMT Databook page no.51]

$K = 200w/mk$

$w = 1.5cm = 0.015m$

$t = 1mm = 0.001m$

$d = 2.5$ cm

$r = \dfrac{2.5}{2} = 1.25$ cm $= 0.0125m$

$T_b = 170°C$

$T_\infty = 25°C$

$H = 130$ w/m²k

To Find

1. Fin efficiency, η
2. Heat loss, Q

Solution

Step 1

$$Q = \eta \, A_s \, h \, (T_b - T_\infty) \quad ....(1)$$

[From HMT Databook page no.51]

$$L_c = L + t/2$$

$$= 0.015 + \frac{0.001}{2}$$

$$L_c = 0.0155\,m$$

$$r_{2c} = r_1 + L_c$$

$$= 0.0125 + 0.0155$$

$$r_{2c} = 0.028\,m$$

$$A_s = 2\pi \left[r_{2c}{}^2 - r_1{}^2 \right]$$

$$= 2\pi \left[(0.028)^2 - (0.0125)^2 \right]$$

$$A_s = 3.944 \times 10^{-3}\,m^2$$

$$A_m = t \left[r_{2c} - r_1 \right] = 0.001\,[0.028 - 0.0125]$$

$$A_m = 1.55 \times 10^{-5}\,m^2$$

Step 2

$$X\ axis = L_C^{1.5} \left(\frac{h}{kAm} \right)^{0.5}$$

$$= (0.0155)^{1.5} \left(\frac{130}{200 \times 1.55 \times 10^{-5}} \right)^{0.5}$$

$$X\ axis = 0.395 \simeq 0.4$$

$$Curve \rightarrow \frac{r_{2C}}{r_1} = \frac{0.028}{0.0125} = 2.24$$

X axis value is 0.4 and curve value is 2.24 corresponding y axis.

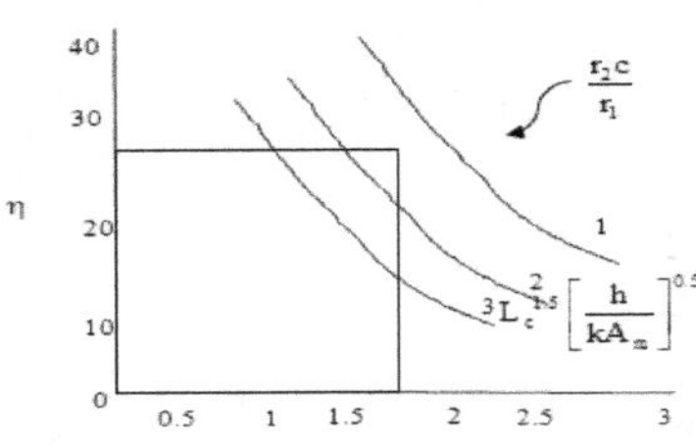

From this graph, we know that

Fin efficiency η = 82%

Step 3

=> $Q = 0.82 \times 3.944 \times 10^{-3} \times 130 \, (170 - 25)$

$Q = 60.96$ w

Result

1. Fin efficiency η = 82%
2. Heat loss, $Q = 60.96$w

Unit 2

Convection

2.1. Convection

Convection is pro-dominant mode of heat transfer in liquid and gas.

Convection is a process of heat transfer that will occur between a solid surface and a fluid medium when they are at different temperatures.

2.1.1. *Newton's Law of Convection*

Heat transfer from the moving fluid to solid surface is given by the equation,

$Q = hA (T_w - T_\infty)$

This equation is referred to as Newton's law of cooling.

Where,

      h   – Local heat transfer coefficient in $w/m^2 k$.

      A   – Surface area in m^2.

      T_w – Surface (or) wall temperature in k.

      T_∞ – Temperature of the fluid ink.

2.1.2. *Types of Convection*

1. Natural or Free convection
2. Forced convection

S.No	Free (or) Natural convection	Forced convection
1	In free convection the motion of fluid is by natural means	In forced convection the motion of fluid is created by some external means.
2	It does not require any external source	It does require external source such as fan, blower
3		
4	Velocity zero	Velocity occurs

2.2. Dimensional Analysis

Dimensional analysis is a mathematical method which makes use of the study of the dimensions for solving several engineering problems.

This method can be applied to all types of fluid resistances, heat flow problems and many other problems in fluid mechanics and thermodynamics.

2.2.1. Dimensions

In dimensional analysis, the various physical quantities used in fluid phenomenon can be expressed in terms of fundamental quantities.

These fundamental quantities are MLtT, where

M - Mass

L - Length

t - Temperature

T - Time

Table 2.1: Dimension

S.No	Quantity	Symbol	Units (SI)	Dimensions (M, L, T, Q, System)
1.	Tube diameter	D	m	L
2.	Fluid density	ρ	Kg/m^3	ML^{-3}
3.	Fluid velocity	U	m/s	LT^{-1}
4.	Fluid viscosity	μ	N-S/m^2	$ML^{-1}T^{-1}$
5.	Specific heat	C_p	J/kg k	$L^2T^{-2}t^{-1}$
6.	Thermal conductivity	K	w/mk	$MLT^{-3}t^{-1}$
7.	Heat transfer co-efficient	h	w/m^{2}k	$MT^{-3}t^{-1}$
8.	Heat	Q	J	ML^2T^{-2}
9.	Heat transfer rate	q	W	ML^2T^{-3}

2.2.2. Buckingham π Theorem

"If there are 'n' variables in a dimensionally homogeneous equation and if these contain 'm' fundamental dimensions, then the variables are arranged into '(n – m)' dimensionless terms. These dimensionless terms are called π terms.

Step 1

Variable	Symbol	Dimension
1. Tube diameter	D	L
2. Fluid density	ρ	ML^{-3}
3. Fluid velocity	U	LT^{-1}
4. Fluid viscosity	μ	$ML^{-1}T^{-1}$
5. Specific heat	Cp	$L^{2}T^{-2}t^{-1}$
6. Thermal conductivity	k	$MLT^{-3}t^{-1}$
7. Heat transfer co-efficient	h	$MT^{-3}t^{-1}$

Step 2

Total number of variables $\quad$ n = 7

Step 3

Fundamental dimensions $\quad$ m = 4

Step 4

πterm

n – m = 7 – 4 = 3 [π term]

$$\pi_1 = (D)^{a_1} (\rho)^{b_1} (\mu)^{c_1} (k)^{d_1} U \qquad (1)$$

$$\pi_2 = (D)^{a_2} (\rho)^{b_2} (\mu)^{c_2} (k)^{d_2} C_p \qquad (2)$$

$$\pi_3 = (D)^{a_3} (\rho)^{b_3} (\mu)^{c_3} (k)^{d_3} h \qquad (3)$$

Step 5

Applying dimensions

$$\pi_1 = M^0 L^0 T^0 t^0 = (L)^{a_1} (ML^{-3})^{b_1} (ML^{-1}T^{-1})^{c_1} (MLT^{-3}t^{-1})^{d_1} , LT^{-1}$$

$$\pi_2 = M^0 L^0 T^0 t^0 = (L)^{a_2} (ML^{-3})^{b_2} (ML^{-1}T^{-1})^{c_2} (MLT^{-3}t^{-1})^{d_2} , L^2 T^{-2} t^{-1}$$

$$\pi_3 = M^0 L^0 T^0 t^0 = (L)^{a_3} (ML^{-3})^{b_3} (ML^{-1}T^{-1})^{c_3} (MLT^{-3}t^{-1})^{d_3} , ML^{-3} t^{-1}$$

Step 6

Equating π_1 term

Equating power, M	Equating power, L	Equating power, T	Equation power, t
$0 = b_1 + c_1 + d_1$	$0 = a_1 - 3b_1 - c_1 + d_1 + 1$	$0 = -c_1 - 3d_1 - 1$	$0 = -d_1$
$0 = b_1 + (-1) + 0$	$0 = a_1 - 3 + 1 + 0 + 1$	$0 = -c_1 - 0 - 1$	$d_1 = 0$
$b_1 = 1$	$a_1 = 1$	$C_1 = -1$	

Equation (1) becomes

$$\pi_1 = D^1 \rho^1 \mu^{-1} k^0 \, U$$

$$\pi_1 = \frac{D\rho}{\mu} \cdot U$$

$$\pi_1 = \frac{UD}{\left(\dfrac{\mu}{\rho}\right)}$$

$$\pi_1 = \frac{UD}{v} = Re \ (\text{Reynolds number})$$

Step 7

Equating π_2 term

Equ. Power of M	Equ. Power of L	Equ. Power of T	Equ. Power of t
$0 = b_2 + c_2 + d_2$	$0 = a_2 - 3b_2 - c_2 + d_2 + 2$	$0 = -c_2 - 3d_2 - 2$	$0 = -d_2 - 1$
$0 = b_2 + 1 - 1$	$0 = a_2 - 3(0) - 1 - 1 + 2$	$C_2 = 3 - 2$	$d_2 = 1$
$b_2 = 0$	$a_2 = 0$	$c_2 = 1$	

Equ (2) becomes

$$\pi_2 = (D)^0 \, (\rho)^0 \, (\mu)^1 \, (k)^{-1} C_p$$

$$\pi_2 = \frac{\mu . C_p}{k}$$

Step 8

Equating π_3 term

Equating Power of M	Equating Power of L	Equating Power of T	Equ. Power of t
$0 = b_3 + c_3 + d_3 + 1$	$0 = a_3 - 3b_3 - c_3 + d_3$	$0 = -c_3 - 3d_3 - 3$	$0 = -d_3 - 1$
$0 = b_3 + 0 - 1 + 1$	$0 = a_3 - 0 - 0 - 1$	$0 = -c_3 - 3(-1) - 3$	$d_3 = -1$
$b_3 = 0$	$a_3 = 1$	$c_3 = 0$	

Equ (3) becomes

$$\pi_3 = (D)^1 \, (P)^0 \, (\mu)^0 \, (k^{-1}) \, h$$

$$\pi_3 = \frac{D.h}{k}$$

2.3. Dimensionaless Number and their Physical Significance

2.3.1. For Forced Convection

1. Nusselt Number (Nu)

Nusselt number gives the information about relative heat transfer by convection through fluid to conduction through same fluid.

$$N_u = \frac{q_{conv}}{q_{cond}} = \frac{hA\,\Delta T}{KA\,\dfrac{\Delta T}{L}} = \frac{h}{k/L}$$

$$N_u = \frac{hL}{k}$$

Where,

h – Heat transfer co-efficient, w/m²k

L – Length, m

k – Thermal conductivity, w/mk.

Physical Significance,

Consider a hot and cold plate separated by a fluid layer of length L.

Fluid is stationary so, the heat transfer through fluid is by conduction only.

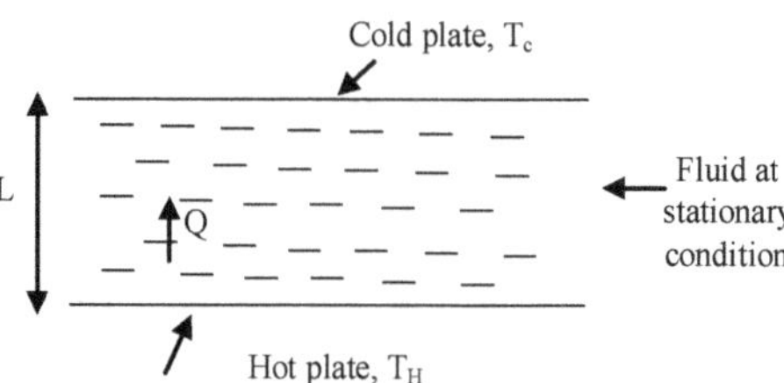

Fig. 2.1

If we impact the motion to fluid then due to bulk motion of fluid now heat is transfer form hot plate to cold plate through same layer of fluid is by convection now.

2. Prandtl Number (Pr)

It is the ratio of the momentum diffusivity to the thermal diffusivity.

$$Pr = \frac{\text{Momentum diffusivity}}{\text{Thermal diffusivity}}$$

$$Pr = \frac{\mu\, Cp}{k} = \frac{v}{\alpha}$$

Where,

v – kinematic viscosity, m^2/s

α - Thermal diffusivity, m^2/s

Significance

Prandtl number provides a measure of the relative effectiveness of the momentum and energy transport by diffusion.

3. Reynolds Number (Re)

It is defined as the ratio of inertia force to viscous force.

$$Re = \frac{Inertia\ force}{Viscous\ force} = \frac{\rho U^2 L^2}{\mu UL} = \frac{\rho UL}{\mu}$$

$$Re = \frac{UL}{v}$$

Where,

U – Velocity, m/s

L – Length, m

$$v = \frac{\mu}{p} = kinematic\ viscosity,\ m^2/s$$

Significance

Reynolds number, is therefore a measure of relative magnitude of the inertia force to the viscous force occurring in the flow.

2.3.2. For Natural Convection

4. Grash of Number (Gr)

It is defined as the ratio of product of inertia force and buoyancy force to the square of viscous force,

$$Gr = \frac{Inertia\ force\ x\ Buoyancy\ force}{(Viscous\ force)^2}$$

$$= \frac{\rho U^2 L^2\ x\ \rho \beta g\ \Delta T L^3}{(\mu\, UL)^2} = \frac{g\rho^2 \beta L^3 \Delta T}{\mu^2}$$

$$Gr = \frac{g \times \beta \times L^3 \times \Delta T}{\nu^2}$$

Where,

β - coefficient of expansion, k^{-1}

L – Length, m

v – kinematic viscosity, m^2/s

ΔT – Temperature difference, k.

Significance

Grash of number has a role in free convection similar to that played by Reynolds number in forced convection.

5. Rayleigh Number (Ra)

$$Ra = Gr.\ Pr = \frac{g\ \beta \Delta T\ L_c^{\ 3}}{\gamma\ \alpha}$$

2.3.3. Laminar Flow

The flow of fluid in highly ordered motion and smoothly is called laminar flow or stream line flow. The flow of high viscosity fluids such as oil at low velocities is typically laminar flow.

2.3.4. Turbulent Flow

Flow of fluid with highly disordered motion is called turbulent flow. The flow of low viscosity fluid at higher velocities is typically turbulent flow.

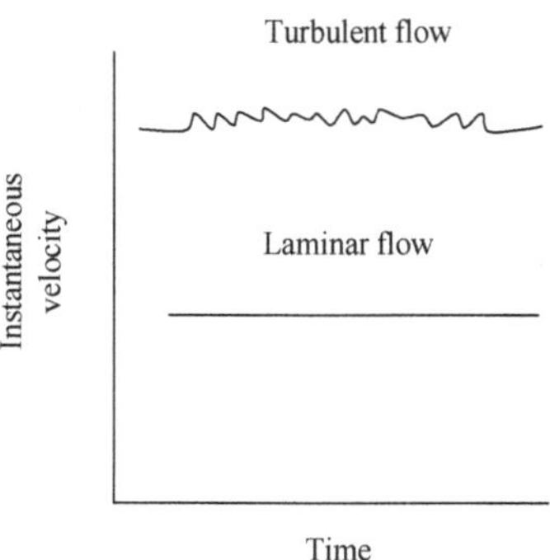

Fig. 2.2

S.NO	DESCRIPTION	FORCE CONVECTION	HMT DB Pg.No	FREE (or) NATURAL CONVECTION	HMT DB Pg.No
1		IF VELOCITY IS GIVEN	-	IF VELOCITY IS NOT GIVEN	-
2	To find film temperature	$$T_f = \frac{Tw + T\infty}{2}$$ Tw – Plate surface temperature ^{0}C T∞ - fluid temperature ^{0}C	113	$$T_f = \frac{Tw + T\infty}{2}$$	135
3	Properties of fluid	From T_f, For water For air (or) gases To take following properties 1) Density in kg/m^3 2) Kinematic viscosity (v) in m^2/s 3) Thermal diffusivity (α) in m^2/s 4) Prantel number (Pr) 5) Specific heat (c) in J/kgk 6) Thermal conductivity (K) in W/mk	22 34	From T_f, For water For air (or) gases To take following properties	22 34
4	To find	Reynolds number For plate $$Re = \frac{UL}{V}$$ Laminar flow Turbulent flow Re < 5 x10^5 Re >5x10^5	112 113	Grashof number i) For vertical plate $$Gr = \frac{g * \beta * L^3 * \Delta T}{v^2}$$ (β) coefficient of thermal expansion in (1/k) For water For air (or) gases from Tf $\Delta T = Tw - T\infty$ Laminar flow turbulent flow Grpr< 10^{9}Grpr> 10^9 ii) For horizontal plate $$Gr = \frac{g * \beta * L_c^3 * \Delta T}{v^2}$$ Lc – characteristic length = W/2 W – width of plate	112 30 135
	For cylinder and sphere	$$Re = \frac{UD}{V}$$ D – diameter in m	112	$$Gr = \frac{g * \beta * D^3 * \Delta T}{v^2}$$ D – diameter in m	112
5	To find Nusselt number (Nu)	For plate laminar flow Nu =0.332 (Re)$^{0.5}$ (Pr)$^{0.333}$ For plate turbulent flow Fully from leading edges. Nu = 0.0296(Re)$^{0.8}$(Pr)$^{0.33}$ For plate turbulent flow (Laminar –turbulent constant wall temperature) Nu=Pr$^{0.333}$[0.037(Re)$^{0.8}$-871]	113 114 115	i) For vertical plate laminar flow Nu = 0.59 (GrPr)$^{0.25}$ 10^4<GrPR<10^9 For vertical plate turbulent flow Nu = 0.10(GrPr)$^{0.333}$ ii) For horizontal plate turbulent flow upper surface heated Nu = 0.54(GrPr)$^{0.25}$ 2x10^4<GrPr<8x10^6 For horizontal plate lower surface heated Nu = 0.27(GrPr)$^{0.25}$ 10^5<GrPr<10^{11}	136 136 137
		For cylinder Nu=C(Re)^m(Pr)$^{0.333}$ To find C and m value from (Re) For sphere Nu = 0.37(Re)$^{0.6}$	116 120	For cylinder (horizontal) Nu=C[GrPr]^m For sphere Nu=2+0.43[GrPr]$^{0.25}$	138 138

S.NO	DESCRIPTION	FORCE CONVECTION	HMT DB Pg.No	FREE (or) NATURAL CONVECTION	HMT DB Pg.No
6	To find local heat transfer co-efficient (h_x)	For plate Nusselt number (Nu) = $h_x l/k$ Average heat transfer Co-efficient Laminar flow(h) = $2*h_x$ Turbulent flow (h) = $1.25 *h_x$ For cylinder and sphere $$Nu = \frac{hD}{K}$$	112	$$Nu = \frac{hL}{K}$$ i) For vertical plate ii) For horizontal plate (upper surface) $$Nu = \frac{h_u L_c}{K}$$ For horizontal plate Lower surface $$Nu = \frac{h_L L_c}{K}$$ For cylinder and sphere $$Nu = \frac{hD}{K}$$	112
7	To find heat transfer	$Q = hA(Tw - T\infty)$ for plate $A = W * L \quad (m^2)$ For cylinder $A = \pi DL \quad (m^2)$ For sphere $A = 4\pi r^2 (m^2)$		i) For vertical plate $$Q = hA(Tw - T\infty)$$ ii) For horizontal plate $$Q = (h_u + h_l)A(Tw - T\infty)$$ iii) For cylinder and sphere $$Q = hA(Tw - T\infty)$$	
		For plate (laminar flow) - Hydrodynamic boundary: Layer thickness (8) $\delta_{hx} = 5*x*(Re)^{-0.5}(m)$ Thermal boundary layer thickness $\delta_{Tx} = \delta_{hx}(Pr^{-0.333})(m)$ Local friction coefficient $C_{fx} = 0.664(Re)^{-0.5}$ Average friction coefficient $\overline{C_{fl}} = 1.328(Re)^{-0.5}$ Shearing stress (or) local Shearing stress (Tx)in N/m² $C_{fx} = \dfrac{\frac{\tau_x}{pu^2}}{2}$ Total dray force on the plate (Fd) $\overline{C_{fl}} = \dfrac{\frac{\tau}{pu^2}}{2}$ Average shear stress = $\tau\left(\dfrac{N}{m^2}\right)$ F_D= area x average shear stress F_D= W * L * T (N) Mass flow rate $\dot{m} = \dfrac{5}{8pu[\delta_{2x} - \delta_{1x}]}(\frac{kg}{s})$ $(\delta_{2x} = \delta_x)(\delta_{1x} = 0)$ For plate turbulent flow boundary layer thickness $\delta = 0.37 * x * (Re)^{-0.2}$ Local skin friction coefficient $C_{fx} = 0.0576(Re)^{-0.2}$ Average friction coefficient $C_{fl} = 0.074(Re)^{-0.2} - 1742(Re)^{-1.0}$	113	Boundary layer thickness $\delta_x = [3.93*(Pr)^{-0.5}(0.952 + Pr)^{0.25}*(G^r)^{-0.25}]*x$ Maximum velocity $u_{max} = 0.766*V*(0.952 + pr)^{-\frac{1}{2}}*[g\beta(Tw - T\infty)/V^2]^{\frac{1}{2}}*x^{\frac{1}{2}}$ Mass flow rate $$m = 1.7 * e * V[Gr/(pr)^2(pr + 0.952)]^{0.25}$$	

2.4. Solved Problem on Flat surfaces Forced Convection

Re <5 x 10^5	Re > 5 x 10^5
Laminar flow	Turbulent flow.

1. Air at 20°C at atmosphere pressure flows over a flat plate at a velocity of 3.5 m/s. If the plate is 0.5m wide and at 60°C, calculate the following at x = 0.400m.

   (i) Boundary layer thickness

   (ii) Local friction coefficient

   (iii) Average friction coefficient

   (iv) Shearing stress due to friction

   (v) Thermal boundary layer thickness

   (vi) Local convective heat transfer co-efficient

   (vii) Average convective heat transfer co-efficient

   (viii) Rate of heat transfer by convection

   (ix) Total drag force on the plate

   (x) Total mass flow rate through the boundary

Given

Fluid temperature, T_∞ = 20°C

Velocity, v = 3.5 m/s

Wide, w = 0.5m

Plate surface temperature, T_w = 60°C

Distance, x = 0.400m

To Find

i. Boundary layer thickness, δ

ii. Local friction, co-efficient, C_{f_x}

iii. Average friction co-efficient, C_{f_L}

iv. Shearing stress due to friction, τ_x

v. Thermal boundary layer thickness, δ_{Tx}

vi. Local convective heat transfer coefficient, h_x

vii. Average convective heat transfer coefficient, h

viii. Rate of heat transfer by convection, Q

ix. Total drag force on the plate, F_D

x. Total mass flow rate through boundary, $\overset{\bullet}{m}$

Solution

Step 1

Velocity is given

Step 2

Film temperature, $T_f = \dfrac{T_w + T_\infty}{2} = \dfrac{60 + 20}{2} = 40°C$

Step 3

Properties of air at 40°C

[From HMT Databook, page no.34]

ρ - 1.128 kg/m³

v - 16.96 x 10⁻⁶ m²/s

P_r - 0.699

K - 0.02756 w/mk

Step 4

Reynolds number, $Re = \dfrac{UX}{v}$

[From HMT Databook page.no. 112]

$$= \dfrac{3.5 \times 0.400}{16.96 \times 10^{-6}}$$

Re $= 8.25 \times 10^4 < 5 \times 10^5$

Since Re $< 5 \times 10^5$, flow is laminar

[Refer HMT Databook. Page. No. 113]

Step 5

(i) Local Heat Transfer Co-efficient, h_x

Local Nusselt Number, $Nu_x = 0.322\,(Re)^{0.5}\,(Pr)^{0.333}$

[From HMT D.b.pg.no.113]

$Nu_x = 0.332\,(8.25 \times 10^4)^{0.5} \times (0.699)^{0.333}$

$Nu_x = 84.7$

Local Nusselt Number, $Nu_x = \dfrac{h_x L}{k}$

$$h_x = \frac{Nu_x \times k}{L} = \frac{84.7 \times 0.02756}{0.400}$$

$[\therefore \ x = L = 0.400\text{m}]$

$h_x = 5.83 \ \text{w/m}^2\text{k}$

Step 6

(ii) Average Heat Transfer Co-efficient, h

$h \quad = 2 \times h_x$

$\quad\quad = 2 \times 5.83$

$h \quad = 11.66 \ \text{w/m}^2\text{k}$

Step 7

(iii) Rate of Heat Transfer, Q

Heat transfer, $Q = hA \ (T_w - T_\infty)$

$\quad\quad\quad = h \times w \times L \times (T_w - T_\infty)$

$\quad\quad\quad = 11.66 \times 0.5 \times 0.400 \times (60 - 20)$

$\quad\quad\quad [\therefore \ x = L = 0.400\text{m}]$

$\quad\quad\quad Q = 93.28 \ \text{w}$

Step 8

[From HMT Data Book Page No. 113]

(iv) Boundary Layer Thickness (or) Hydrodynamic

Boundary layer thickness, δ or δ_h at

$X = 0.400\text{m}$

$\delta \text{or} \ \delta_{h_x} \quad = 5 \times x \times (Re)^{-0.5}$

$\quad\quad\quad\quad = 5 \times 0.400 \times (8.25 \times 10^4)^{-0.5}$

$\delta_x \quad\quad\quad = 6.96 \times 10^{-3}\text{m}$

Step 9

(v) Thermal Boundary Layer Thickness δ_{T_x}

$\delta_{T_x} = \delta_{T_x} \times (Pr)^{-0.333} = 6.96 \times 10^{-3} \times (0.699)^{-0.333}$

$\delta_{T_x} = 7.84 \times 10^{-3} \ \text{m}$

Step 10

(vi) Local Friction Co-efficient, C_{fx}

$$C_{fx} = 0.664 \, (Re)^{-0.5}$$

$$= 0.664 \, (8.25 \times 10^{-4})^{-0.5}$$

$$C_{fx} = 2.31 \times 10^{-3}$$

Step 11

(vii) Average Friction Co-efficient, $\overline{CfL}$

$$\overline{CfL} = 1.328 \, (Re)^{-0.5}$$

$$= 1.328 \, (8.25 \times 10^{4})^{-0.5}$$

$$\overline{CfL} = 4.623 \times 10^{-3}$$

Step 12

(viii) Shearing Stress (or) Local Shear Stress, τ_x

$$C_{fx} = \frac{\tau_x}{\rho V^2 / 2}$$

$$\tau_x = C_{fx} \times \frac{\rho V^2}{2} = 2.31 \times 10^{-3} \times \frac{1.128 \times (3.5)^2}{2}$$

$$\tau_x = 0.0159 \, N/m^2$$

Step 13

(ix) Total Drag Force on the Plate F_D

$$C_{fx} = \frac{\tau_x}{\rho V^2 / 2}$$

$$\tau_x = C_{f_L} \times \frac{\rho V^2}{2} = 4.623 \times 10^{-3} \times \frac{1.128 \times (3.5)^2}{2}$$

Average Shear Stress, $\tau = 0.0319 \, N/m^2$

Drag force, F_D = Area x Average Shear Stress

$$= w \times L \times \tau = 0.5 \times 0.4 \times 0.0319$$

$$F_D = 6.38 \times 10^{-3} \, N$$

Drag Force on Two Sides of the Plate= $2 \times 6.38 \times 10^{-3}$

Total drag of force, $F_D = 0.0127$N.

Step 14

(x) Total Mass Flow Rate, $m^\bullet$

$$m^\bullet = \frac{5}{8}\rho V [\delta_{2x} - \delta_{1x}]$$

Here, $\delta_{1x} = 0$, $\delta_{2x} = \delta_x = 6.96 \times 10^{-3}$m

$$m^\bullet = \frac{5}{8} \times 1.128 \times 3.5 \times [6.96 \times 10^{-3}] => m^\bullet = 0.017 \text{ kg/s}$$

Result

  i. δ or $\delta_{hx} = 6.96 \times 10^{-3}$ m

 ii. $C_{fx} = 2.31 \times 10^{-3}$

iii. $\overline{CfL} = 4.623 \times 10^{-3}$

 iv. $\tau_x = 0.0159$ N/m^2

  v. $\delta_{Tx} = 7.84 \times 10^{-3}$m

 vi. $h_x = 5.83$ w/m^{2}k

vii. $h = 11.66$ w/m^{2}k

viii. $Q = 93.28$w

 ix. $F_D = 0.0127$N

  x. $m^\bullet = 0.017$ kg/s

2. Air at a pressure of 8 KN/m^2 and a temperature of 250°C flows over a flat plat 0.3m wide and 1m long at a velocity of 8 m/s. If the plate is to be maintained at a temperature of 78°C, estimate the rate of heat to be removed continuously from the plate.

Given

Pressure, $P = 8$ KN/m$^2 = 8 \times 10^3$ N/m^2

Fluid temperature, $T_\infty = 250^o$C

Wide, $w = 0.3$m

Length, $L = 1$m

Velocity, $v = 8$ m/s

Plate temperature, $T_w = 78^o$C

To Find

Heat transfer

Solution

Step 1

Velocity is given

$$\text{Film temperature, } T_f = \frac{T_w + T_\infty}{2}$$

$$= \frac{78 + 250}{2}$$

$$T_f = 164\,^oC$$

Step 2

Properties of air at 164°C

[From HMT Data book page no. 34]

$\rho = 0.810$ kg/m³

$v = 30.08 \times 10^{-6}$ m²/s

$K = 0.03645$ w/mk

$P_r = 0.682$

Note: Given pressure is not atmosphere pressure so, kinematic viscosity will vary with pressure (Pr, K, Cp are same for all pressure)

$$\text{Kinematic viscosity, } v = v_{atm} \times \frac{P_{atm}}{P_{given}}$$

$$= 30.08 \times 10^{-6} \times \frac{1 \times 10^{5} \ N/m^2}{8 \times 10^{3} \ N/m^2}$$

$$V = 3.76 \times 10^{-4} \text{ m}^2/\text{s}$$

Step 3

$$\text{Reynolds number } Re = \frac{UL}{v} = \frac{8 \times 1}{3.76 \times 10^{-4}}$$

$$Re = 2.1 \times 10^{4} < 5 \times 10^{5}$$

Since $Re < 5 \times 10^{5}$ flow is laminar

Step 4

Forflat plate, laminar flow

[From HMT Databook page.no. 113]

Local Nusselt number $(Nu)_x$ $= 0.322 \times (Re)^{0.5} (Pr)^{0.333}$

$= 0.322 \times (2.1 \times 10^4)^{0.5} \times (0.682)^{0.333}$

$Nu_x = 42.35$

We know that

$$Nu_x = \frac{h_x L}{k} \qquad \text{[From HMT Pg.No.112]}$$

$$42.35 = \frac{h_x \times 1}{0.03645}$$

Local heat transfer co-efficient $h_x = 1.54 \ w/m^2k$.

Step 5

Average heat transfer co-efficient, $h = 2 \times h_x$

$h = 2 \times 1.54$

$h = 3.08 \ w/m^2k$

Step 6

Heat transfer, Q $= h \times A \times (T_\infty - T_w)$

$= h \times W \times L \times (T_\infty - T_w) = 3.08 \times 1 \times 0.3 \times (250 - 78)$

Q $= 158.9 \ w$

Heat transfer both sides of the plate

Q $= 2 \times 158.9$

Q $= 317.88 \ w$

Result

Heat transfer, Q = 317.85 w

3. Air at 20°C and one atmosphere flows over a flat plate at 35 m/s. The plate is 75 cm long and is maintained at 60°C. Calculate the heat transfer per unit width of the plate. Also calculate the turbulent boundary layer thickness at end of the plate assuming it to develop from the leading edge of the plate and local skin friction co-efficient.

Given

Fluid temperature, $T_\infty = 20°C$

Velocity, U = 35 m/s

Length, L = 75 cm = 0.75m

Plate temperature, $T_w = 60°C$

Width, w = 1m

To Find

1. Heat transfer
2. Boundary layer thickeness
3. Local skin friction co-efficient

Solution

Step 1

Velocity is given

Film temperature, $T_f = \dfrac{T_w + T_\infty}{2} = \dfrac{60 + 20}{2}$

$T_f \quad = 40^\circ C$

Step 2

Properties of air at $40^\circ C$

$\rho = 1.128 \ kg/m^3$

$K = 0.02756 \ w/mk$

$v = 16.96 \times 10^{-6} \ m^2/s$

$Pr = 0.699$

Step 3

Reynolds Number, $Re = \dfrac{UL}{v} = \dfrac{35 \times 0.75}{16.96 \times 10^{-6}}$

$$Re = 1.54 \times 10^6 > 5 \times 10^5$$

Since $Re > 5 \times 10^5$, flow is turbulent

Step 4

For Flat Plate, turbulent flow [Fully turbulent given]

[From HMT Data book page no. 114]

$Nu_x = 0.0296 \ (Re)^{0.8} \times (Pr)^{0.333} = 0.0296 \ (1.54 \times 10^6)^{0.8} \times (0.699)^{0.333}$

$Nu_x = 2341.6$

Local Nusslet Number $Nu_x = \dfrac{h_x \times L}{k}$

$2341.6 = \dfrac{h_x \times 0.75}{0.02756}$

$h_x = 86.04 \ w/m^2 k$

Step 5

Average heat transfer co-efficient $h = 1.25 h_x$

$h = 1.25 \times 86.04$

$h = 10.7.55 \ w/m^2k$

Step 6

(i) Heat Transfer $Q = h \times A \times (T_w - T_\infty)$

$$= h \times w \times L \times (T_w - T_\infty) = 107.55 \times 0.75 \times 1 \times (60 - 20)$$

$$Q = 3226.50 \ w$$

Step 7

(ii) Boundary Layer Thickness δ

$\delta \quad = 0.37 \times xx \ (Re)^{-0.2} = 0.37 \times 0.75 \times (1.54 \times 10^6)^{-0.2}$

$\delta \quad = 0.0160 m$

Step 8

(iii) Local Skin Friction Co-efficient

$Cfx \ = 0.0576 \ (Re)^{-0.2} = 0.0576 \ (1.54 \times 10^6)^{-0.2}$

$Cfx \ = 0.4434$

Result

1. Heat transfer $Q = 3226.50 \ w$
2. Boundary layer thickness $\delta = 0.0160 \ m$
3. Local skin friction co-efficient $Cfx = 0.4434$

4. Atmosphere air at 275k and a free stream velocity of 20 m/s flows over a flat plate 1.5m long that is maintained at a uniform temperature of 325 k. calculate the average heat transfer co-efficient over the region where the boundary layer is laminar, the average heat transfer co-efficient over the entire length of the plate and the total heat transfer rate from the plate to the air over the length 1.5m and width 1m. Assume transition occurs at $Re = 2 \times 10^5$

Given

Fluid temperature, $T_\infty = 275k = 2°C$

Velocity, $U = 20 \ m/s$

Length, $L = 1.5m$

Plate surface temperature, Tw = 325 k = 52°C

Width, w = 1m

Critical Reynolds number, Re_c = 2 x 10^5

To Find

1. Average heat transfer co-efficient, h_l
   [Boundary layer laminar]
2. Average heat transfer co-efficient, h_t
   [Entire length of the plate]
3. Total heat transfer rate, Q

Solution

Step 1

Velocity is given

Film temperature, $\quad T_f = \dfrac{T_w + T_\infty}{2} = \dfrac{52 + 2}{2}$

$$T_f = 27°C$$

Step 2

Properties of air at 27°C ≈ 25°C

[From HMT Databook page no. 34]

ρ = 1.185 kg/m^2

v = 15.53 x 10^{-6} m^2/s

Pr = 0.702

K = 0.02634 w/mk

Step 3

Case (i)

Reynolds number, Re = $\dfrac{UL}{v}$

Transition occurs at Re_c = 2 x 10^5

i.e., flow is laminar up to Reynolds number value is 2 x 10 after that flow is turbulent.

$$=> 2 \times 10^5 = \frac{20 \times L}{15.53 \times 10^{-6}}$$

=> L = 0.155m

Step 4

For flat plate, laminar flow

Local Nusselt number, $Nu_x = 0.332 \, (Re)^{0.5} \, (Pr)^{0.333}$

[From HMT data book page no. 113]

$\Rightarrow Nu_x = 0.332 \, (2 \times 10^5)^{0.5} \, (0.702)^{0.333}$

$\Rightarrow Nu_x = 131.97$

We know that,

$$Nu_x = \frac{h_x L}{k}$$

$$131.97 = \frac{h_x \times 0.155}{0.02634}$$

$\Rightarrow h_x = 22.42 \text{ w/m}^2\text{k}$

Local heat transfer co-efficient, $h_x = 22.42 \text{ w/m}^2\text{k}$

Step 5

Average heat transfer co-efficient, $h = 2 \times h_x = 2 \times 2\,2.42$

$h \quad = 44.84 \text{ w/m}^2\text{k}$

Step 6

Case (ii)

Reynolds number, $Re_L = \dfrac{UL}{\nu}$

[For entire length]

$$= \frac{20 \times 1.5}{15.53 \times 10^{-6}}$$

$Re_L = 1.93 \times 10^6 > 5 \times 10^5$

Since $Re_L > 5 \times 10^5$, flow is turbulent.

Step 6

For flat plate, laminar – turbulent combined flow.

Average Nusselt number, (Nu)

$Nu = (Pr)^{0.333} \, [0.037 \, (Re_L)^{0.8} - 871]$

$Nu = (0.702)^{0.333} \, [0.037 \, (1.93 \times 10^6)^{0.8} - 871]$

$Nu = 2737.18$

We know that

Nusslet number, $Nu = \dfrac{hL}{k}$

$\Rightarrow 2737.18 = \dfrac{h \times 1.5}{0.02634}$

$\Rightarrow h = 48.06 \ w/m^2 k$

Average heat transfer co-efficient for turbulent flow, $h_t = 48.06 \ w/m^2 k$

Step 7

Total heat transfer rate, Q $\quad = h_t A \Delta T$

$\quad\quad = h_t \times w \times L \times (T_w - T_\infty) = 48.06 \times 1 \times 1.5 \times (52 - 2)$

Q $\quad = 3604.5 \ w$

Result

1. Average heat transfer co-efficient
   [Boundary layer is laminar]
   $h_l = 44.84 \ w/m^2 k$
2. Average heat transfer co-efficient
   [For entire length of the plate]
   $h_t = 48.06 \ w/m^2 k$
3. Total heat transfer rate, Q = 3604.5 w

2.5. Solved Problems – Flow Over Cylinder

1. Air at 15°C, 30 km/h flows over a cylinder of 400mm Diameter and 1500mm height with surface temperature of 45°C calculate the heat loss

Given

Fluid temperature, $T_\infty = 15°C$

Velocity, U $\quad = 30 \ km/h$

$\quad\quad = \dfrac{30 \times 10^3}{3600} \ m/s$

U $\quad = 8.33 \ m/s$

Diameter, D = 400 mm = 0.4m

Length, L = 1500mm = 1.5m

Plate surface temperature, $T_w = 45°C$

To Find

Heat loss

Solution

Step 1

Velocity is given

Film temperature, $\quad T_f = \dfrac{T_w + T_\infty}{2} = \dfrac{45+15}{2}$

$$T_f = 30^\circ C$$

Step 2

Properties of air at $30^\circ C$

[From HMT data book page no. 34]

$\rho = 1.165 \text{ kg/m}^3$

$v = 16 \times 10^{-6} \text{ m}^2/\text{s}$

$Pr = 0.701$

$K = 0.02675 \text{ w/mk}$

Step 3

Reynolds Number, $Re = \dfrac{UD}{v} = \dfrac{8.33 \times 0.4}{16 \times 10^{-6}}$

$$Re_D \quad = 2.08 \times 10^5$$

Step 4

For cylinder,

Nusselt Number, $Nu = C\,(Re)^m (Pr)^{0.333}$

[From HMT Databook page no. 116]

Re_D value is 2.08×10^5, corresponding (value is 0.0266 and m value is 0.805.

$\Rightarrow Nu = 0.0266 \times (2.08 \times 10^5)^{0.805} \times (0.701)^{0.333}$

$Nu = 451.3$

Nusslet Number, $Nu = \dfrac{hD}{k} \Rightarrow 451.3 = \dfrac{h \times 0.4}{0.02675}$

$h = 30.18 \text{ w/m}^2\text{k}$

Heat transfer co-efficient, $h = 30.18 \text{ w/m}^2\text{k}$.

Step 5

Heat transfer $Q = hA_x (T_w - T_\infty)$

$$= h \times \pi \times D \times L \times (T_w - T_\infty)$$

$[\because A = \pi DL]$ $= 30.18 \times \pi \times 0.4 \times 1.5 \times (45 - 15)$

Q $= 1706.6$ w

Result

Heat loss, $Q = 1706.6$ w

2.6.   Solved Problems On Flow Over Spheres

1. Air at 30°C, 0.2 m/s flows across a 120w electric bulb at 130°C. Find the heat transfer and power lost due to convection if bulb diameter is 70mm.

Given:

Fluid temperature, T_∞ $= 30$°C

Velocity, U $= 0.2$ m/s

Heat energy, Q $= 120$w

Surface temperature, T_w $= 130$°C

Diameter, D = 70mm $= 0.070$m

To Find

1. Heat transfer

2. Power lost due to convection

Solution

Step 1

Velocity is given

Film temperature, T_f $= \dfrac{T_w + T_\infty}{2} = \dfrac{130 + 130}{2}$

T_f $= 80$°C

Step 2

Properties of air at 80°C

[From HMT Data book page no. 34]

$\rho = 1$ kg/m^3

$v = 21.09 \times 10^{-6} \; m^2/s$

$Pr = 0.692$

$K = 0.03047 \; w/mk$

Step 3

Reynolds Number, $\quad Re = \dfrac{UD}{v} = \dfrac{0.2 \times 0.070}{21.09 \times 10^{-6}}$

$$Re = 663.82$$

Step 4

For Sphere,

Nusslet Number, $Nu = 0.37 \, (Re)^{0.6}$

[From HMT Data book page no. 120]

$Nu = 0.37 \, (663.82)^{0.6}$

$Nu = 18.25$

Nusslet Number, $Nu = \dfrac{hD}{k}$

$\Rightarrow 18.25 = \dfrac{h \times 0.070}{0.03047}$

$\Rightarrow h = 7.94 \; w/m^2k$

Heat transfer co-efficient, $h = 7.94 \; w/m^2k$

Step 5

Heat transfer, $Q_2 = hA \, (T_w - T_\infty) = h \times 4\pi r^2 \times (T_w - T_\infty)$

$[\because A = 4\pi r^2] \; = 7.94 \times 4 \times \pi x \left(\dfrac{0.070}{2}\right)^2 (130 - 30)$

$Q_2 \quad = 12.22 \; w$

2. % of heat lost $= \dfrac{Q_2}{Q_1} \times 100 = \dfrac{12.22}{120} \times 100$

$$= 10.18\%$$

Result

1. Heat transfer = 12.22 w

2. Percentage of heat lost = 10.18%

2.7. Concept of Velocity and Thermal Boundary Layer

- Consider parallel flow of fluid over a flat plate as shown in fig.2.3. Surface is assumed as flat plate.
- The x co-ordinate is measured along plate surface y is measured from surface in normal direction to it.
- Fluid approaches plate at uniform velocity of U_∞ which is called free stream velocity.
- The velocity of fluid layer adjacent to plate is zero because of no slip condition.

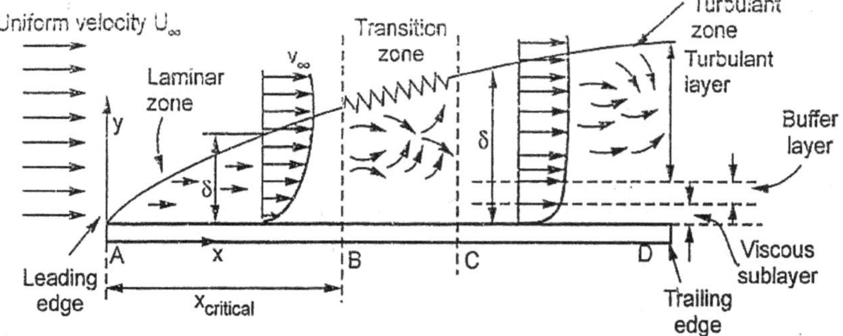

Fig. 2.3: Concept of Velocity Boundary Layer

- This motionless layer tries to slow down the particles of neighbouring fluid layer as a result of friction between the adjoining fluid layers at different velocity.
- This phenomenon continuous in fluid in y direction, thus presence of plate is felt up to some distance δ form plate beyond which free stream velocity U_∞ remains unchanged.
- Due to this velocity component of fluid u, varies from 0 at y = 0 to nearly u = 0.99 U∞ at y = δ. So the region of flow above plate bounded by δ in which viscous effect is predominant is called velocity boundary layer thickness.

Initially form leading edge of plate upon $x_{critical}$ the flow of fluid is well organised and smooth so this region is called laminar zone. i.e. zone AB. In zone BC the fluid mixing is intense and fluid flow becomes random. This region is called transition zone beyond point 'C' the fluid motion becomes highly disorganised and this region is called turbulent boundary layer.

Changes from laminar to turbulent is given by Reynold number,

$$Re = \frac{\rho U\infty X}{\mu}$$

For plate when Re $<5\times10^5$ the flow is said to be laminar.

2.7.1. *Thermal Boundary Layer*

- Consider flow of fluid at uniform temperature of T_∞ over a flat plate at temperature T_s such that $[T_s > T_\infty]$.
- Fluid layer adjacent to surface will reach thermal equilibrium with plate as at same temperature of plate i.e. T_s.
- Fluid layer exchange the energy with adjacent layer as result of which temperature profile develops in flow field that ranges from T_s at surface to T_∞ sufficiently far away of surface.

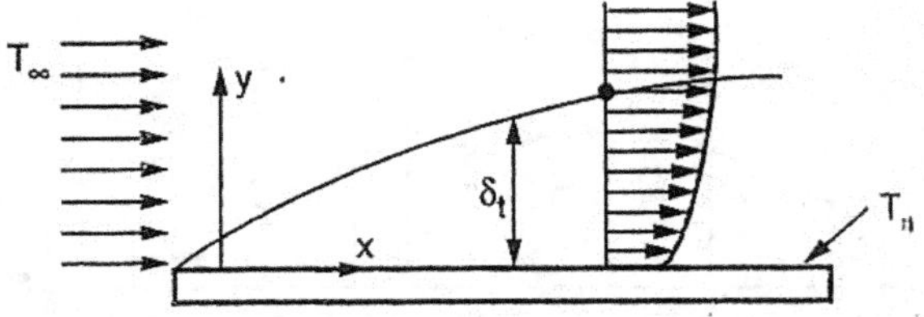

Fig. 2.4: Concept of Thermal Boundary Layer

The flow region over surface in which temperature variation in direction normal to surface is significant is called thermal boundary layer region and the distance measured normal to surface at which $(T - T_s) = 0.99\ (T_\infty - T_s)$ is called thermal boundary layer thickness δ_1.

Thickness of thermal boundary layer increases from leading edge.

The relation between boundary layer and convective heat transfer coefficient may be stated as follows:

At any distance x from leading edge the heat flux may be obtained by applying Fourier law to fluid at y = 0.

$$Q_x = -k_{fluid} \left(\frac{\partial T}{\partial y} \right)_{y=0}$$

Since at surface there is no fluid motion and energy transfer occurs by conduction.

From first layer of fluid at surface to next layer, heat is transferred by convection, so by Newton's law of cooling,

$$Q_x = h_x\ (T_\infty - T_s)$$

From above two equations, we have

$$Hx = \frac{-K_{fluid}\ (\partial T / \partial y)_{y=0}}{(T_s - T_\infty)}$$

Thus conditions in thermal boundary layer which strongly influences the wall temperature gradient $\left(\dfrac{\partial T}{\partial y}\right)_{y=0}$, determine the rate of heat transfer across the boundary layer. Since (T_∞-T_s) is a constant and independent of x, while δ_1 increases with increasing x, thus temperature gradients in thermal boundary layer must decrease with increasing x and it follows that q_x and h_x decreases with increase in x.

2.7.2. *Significance of Boundary Layer*

- When there is a flow of fluid over any surface, there will always exist a velocity boundary layer and hence surface friction.
- Also a thermal boundary layer and hence convection heat transfer will always exist if surface and free stream fluid temperature differs.
- Velocity boundary layer thickness δ is characterized by presence of velocity gradient and shear stress.
- The thermal boundary layer thickness δ_t is characterized by temperature gradient and heat transfer.
- Velocity and thermal boundary layer rarely grows at same rate, and values of δ and δ_t at given location are not same.
- For the engineer the principal manifestations of two boundary layers are respectively surface friction and convection heat transfer. Thus boundary layer parameters are the friction coefficient C_f and heat transfer coefficient 'h' respectively.

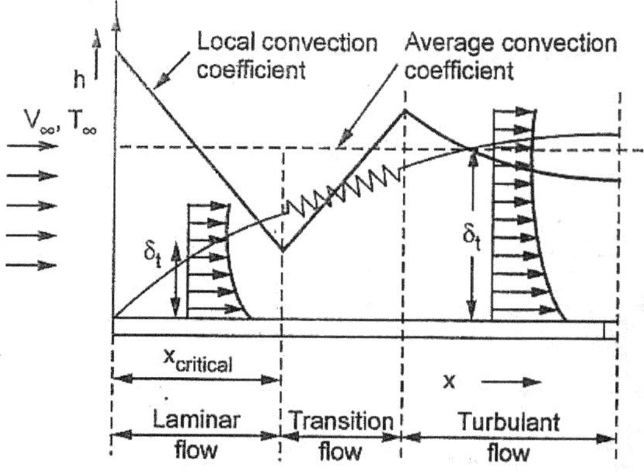

Fig. 2.5: Effect of Thermal Boundary Layer Thickness Over Average and Local Convection Coefficient

From fig. 2.5, we can conclude that the value of h_x at leading edge is maximum because thermal boundary layer thickness δ_t is zero at leading edge. In laminar region h_x decreases as δ_t increases.

In transition flow region h_x increases due to intense mixing of fluid particles.

Again in turbulent region h_x decreases as value of δt increases.

2.8. Internal Forced Convection

When fluid is flowing through the tubes, pipes or duct it is termed as internal forced flow. Generally flow section of circular cross section is called as a pope while flow section with noncircular cross section is called as duct, small cross section pipes are referred as tubes.

- We preferred the circular pipes to transport the liquids as circular pipes can withstand large pressure difference without undergoing large distortion.

- Non circular pipes. i.e. ducts are used to transmit the gases where the pressure difference is very small. Circular tubes gives more heat transfer for least pressure drop at given surface area.

- In the internal forces flow fluid velocity changes from zero at wall to maximum at centre of pipe. So we take average velocity of fluid in the calculation of mass flow rate of fluid. Average velocity remains constant for incompressible flow when cross section area of tube is constant.

- When fluid flows through the pipe due to no slip condition velocity boundary layer develops inside pipe. Similarly when fluid temperature is different than the pipe surface temperature layer will develop inside it.

2.8.1. *Velocity Boundary Layer Development in Circular Pipe For Laminar Flow*

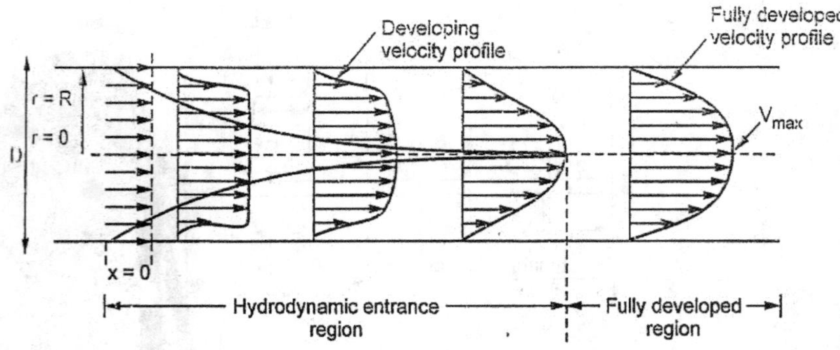

Fig. 2.6: Velocity Boundary Layer Development for Laminar Flow through Pipes

- As shown in the figure 2.6, fluid enters at uniform velocity 'V' into circular pipe. It has uniform velocity distribution at entrance.
- As fluid moves along the pipe fluid layer in contact with surface tends to become stationary due to no slip condition.
- Stationary fluid tries to slow down the adjacent fluid layer and so on. Thus boundary layer develops.
- Distance from pipe surface upto centre of pipe where boundary layer develops is called boundary layer thickness.
- For laminar flow through pipe the shape of fully developed velocity profile is parabolic.

At $r = R$, $V = 0$ and $r = 0$, $V = V_{max}$

At $r = R$, $\tau = \tau_{max}$, $r = 0$, $\tau \cong 0$

As seen in fig. 2.7. After certain distance from entry of pipe velocity profile velocity profile is not changing with distance and flow is called fully developed flow.

Axial distance covered by fluid from entrance upto point of fully developed velocity profile is called entrance length, Le.

In case of fully developed laminar flow,

Le $= 0.0575$ (Re × D)

$$Re = \frac{\rho V_m D}{\mu} \quad V_m = V_{avg}$$

$$V_m = \frac{V_{max}}{2}$$

Local velocity at any radius r,

$$\frac{V}{V_{max}} = 1 - \left(\frac{r}{R}\right)^2$$

r = Radius at any point from centre of pipe

R = outer radius of pipe

Friction factor, $f = \dfrac{16}{Re}$

Pressure drop, $\Delta p = \dfrac{4fL\rho V^2}{2D}$

2.8.2. *Turbulant Flow through Pipe*

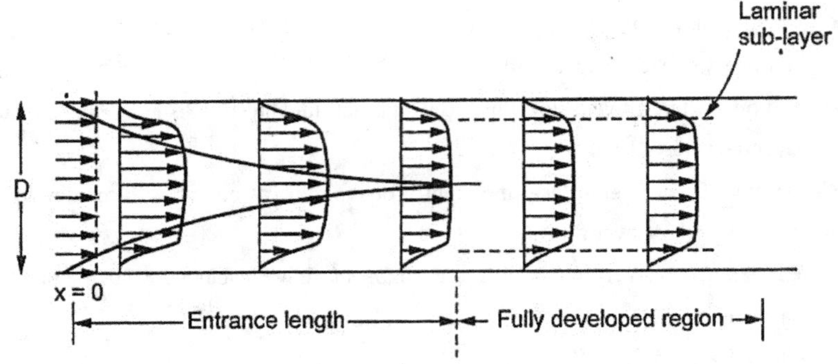

Fig. 2.7: Turbulant Flow through Pipe

In turbulent flow velocity profile is quickly stabilized. The entrance length is short due to large eddies formation.

Velocity profile in turbulant flow is flat in core region.

The hydrodynamic and thermal entry length for turbulant flow is same due to intense mixing during random flucturations. L_h = Entry length $\cong$ 10 D.

2.8.3. *Thermal Boundary Layer in Circular Pipes*

Thermal boundary layer develops in circular pipe when the temperature of fluid is different than the wall temperature of pipe.

When wall is heated two condition can be exist either constant wall surface temperature or constant surface heat flux condition.

Constant surface wall temperature condition:

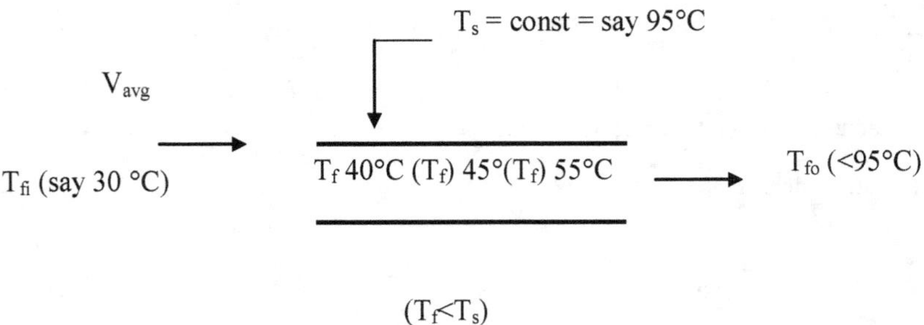

Fig. 2.8: Constant Surface Wall Temperature

In this case of constant surface wall temperature, temperature of surface of tube remains constant at every location.

Temperature of fluid say inlet temperature is T_{fi} increases as fluid passes through tube as fluid is colder than wall and reverse case is happened that is temperature of fluid decreases if $T_f>T_s$.

The thermal boundary layer development and its effect on local heat transfer coefficient is shown in fig.2.9.

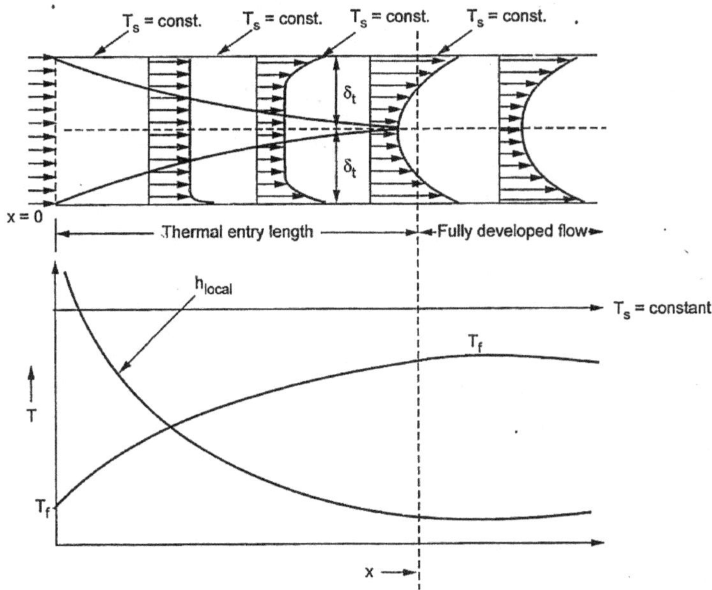

Fig. 2.9: Thermal Boundary Layer Development and its Effect on Local Heat Transfer Coefficient

From fig.2.10 it can be seen that thickness of thermal boundary layer increases with the distance x and after some distance x From fig.2.9 it can be seen that thickness of thermal boundary layer increases with the distance x and after some distance 'x' thermal boundary layer dose not changes with distance which is called fully developed region and entrance length is called thermal entry length, Ln.

Also at entry as thermal boundary layer thickness δ_t is zero value of local heat transfer coefficient is maximum and it decreases as δ_t increases and in fully developed region it remains unchanged.

Fig. 2.10 shows the thermal entry length and hydrodynamic entry length comparison for fluid flow in tube whose Prandtl no.is greater than 1.

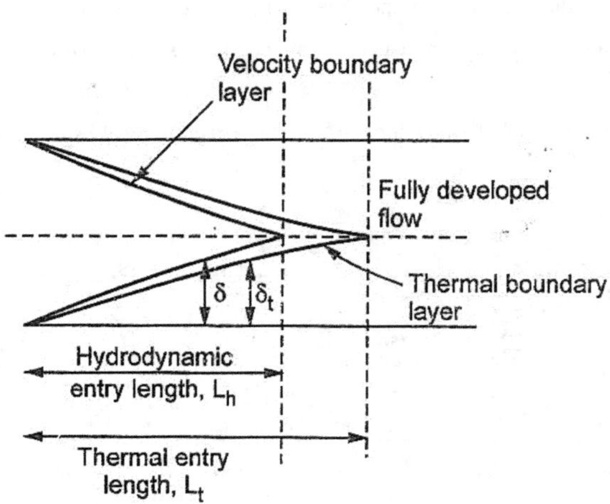

Fig. 2.10: Comparison of Thermal Boundary Layer and Velocity Boundary Layer for Fluid Having Pr>1

From above fig. 2.10 it can be seen that for fluid having Pr>1 such as oils velocity boundary layer thickness is greater than thermal boundary layer thickness.

For fluids such as gas Pr = 1 so $\delta = \delta t$ and for liguid metals Pr<1 $\delta t > \delta$ i.e. Thermal entry length is smaller than hydrodynamic entry length.

Constant Surface Heat Flux Condition

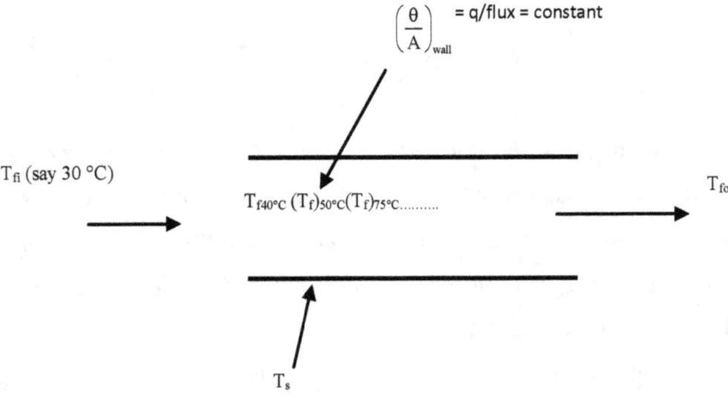

Fig. 2.11: Constant Surface Heat Flux Condition

2.8.4. *Hydraulic Diameter (D_h)*

- To find the heat transfer rate for the fluid flow through pipe or duct of non circular cross section. We have to take hydraulic diameter, D_n,

$$D_h = \frac{4 \text{x} \, C/S \, \text{area of flow}}{\text{wetted perimeter}}$$

For circular pipe, $D_h = \dfrac{4 \text{x} \dfrac{\pi}{4} D^2}{\pi D}$

For rectangular C/S duct of length L and width W,

$$D_h = \frac{4(L \text{x} W)}{(L+W)} = \frac{2(L \text{x} W)}{(L+W)}$$

For Fluid Flowing through Annular Space between Two Circular Pipes

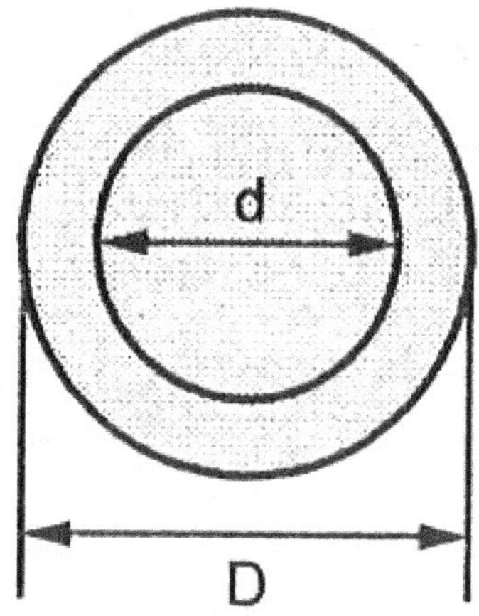

$$D_h = \frac{4 \text{x} \dfrac{\pi}{4} \left[D^2 - d^2 \right]}{\left(\pi D + \pi d \right)}$$

$$= \frac{(D+d)(D-d)}{(D+d)}$$

$D_h = D - d$

Where D = outside pipe diameter

      d = inside pipe diameter

2.9. Forced Convection

If the fluid motion is artificially created by means of an external force like a blower or fan, that type of heat transfer known as forced convection.

2.9.1. *The Local and Average Heat Transfer Coefficients for Plate–Laminar Flow*

Step 1

At the surface of the plate, heat flow may be written as

$$q = \frac{Q}{A} = h_x (T_w - T_\infty)$$

$$= k \left(\frac{\partial T}{\partial y} \right)_{y=0} \qquad\qquad ....(2.1)$$

We know that,

$$= \left(\frac{\partial T}{\partial y} \right)_{y=0} = T_w - T_\infty x \sqrt{\frac{U}{vx}} \times \left(\frac{\partial \theta}{\partial \eta} \right)_{\eta=0}$$

$$= T_w - T_\infty x \sqrt{\frac{U}{vx}} \, x0.332(Pr)^{0.333}$$

$$\left[\because \left(\frac{\partial \theta}{\partial \eta} \right)_{\eta=0} = 0.330(Pr)^{0.333} \right.$$

$$= T_w - T_\infty x \sqrt{\frac{U}{v}} \; x \; x^{\frac{-1}{2}} \; x \, 0.332 x (Pr)^{0.333}$$

$$= (T_w - T_\infty) x \sqrt{\frac{U}{v}} \; x \; x^{\frac{-1}{2}} x \, x^{-1} \, x0.332(Pr)^{0.333}$$

$$= T_w - T_\infty x \sqrt{\frac{Ux}{v}} \frac{1}{x} x0.332x(Pr)^{0.333}$$

$$\left(\frac{\partial T}{\partial y} \right)_{y=0} = \frac{T_w - T_\infty}{x} x \sqrt{\frac{Ux}{v}} \, x0.332(Pr)^{0.333}$$

Substituting $= \left(\dfrac{\partial T}{\partial y} \right)_{y=0}$ in equation (2.1).

$$2.1 \Rightarrow \frac{Q}{A} = h_x(T_w - T_\infty) = k\left(\frac{\partial T}{\partial y}\right)_{y=0}$$

$$\frac{Q}{A} = k\frac{(T_w - T_\infty)}{x} x\sqrt{\frac{Ux}{\nu}}x0.332x(Pr)^{0.333}$$

$$\frac{Q}{A} = 0.332\,x\,\frac{K}{x}\,x\,(T_w - T_\infty)\left(Re\right)^{\frac{1}{2}}x\left(\rho_r\right)^{0.333}$$

$$\left[\therefore Re = \frac{Ux}{\nu}\right]$$

$h_x(T_w\text{-}T\infty) = 0.332\,\dfrac{k}{x}\,(T_w\text{-}T\infty)\,x\,(Re)^{0.5}\,(Pr)^{0.333}$

$h_x = 0.332\,x\,\dfrac{k}{x}\,x(Re)^{0.5}\,(Pr)^{0.333}$

Local heat transfer coefficient, $h_x = 0.332\,x\,(Re)^{0.5}\,(Pr)^{0.333}$ ....(2.2)

We know that,

Local Nusselt number, $Nu_x \quad = \dfrac{h_x x}{k}$

$$Nu_x = \frac{0.332x\dfrac{k}{x}(Re)^{0.5}(Pr)0.333x\,x}{k}$$

Local heat transfer coefficient, $h_x \qquad 0.332\,x\,(Re)^{0.5}\,(Pr)^{0.333}$ ....(2.3)

Step 2

The average heat transfer coefficient, h is given by

$$h \quad = \frac{1}{L}\int_0^L h_x d_x$$

$$= \frac{1}{L}\int_0^L 0.332x\frac{k}{x}(Re)^{0.5}(Pr)^{0.333}\,dx$$

$$= \frac{1}{L}\int_0^L 0.332x\frac{k}{x}\left(\frac{Ux}{\nu}\right)^{0.5}(Pr)^{0.333}\,dx$$

$$\left[\therefore Re = \frac{Ux}{\nu}\right]$$

$$= \frac{1}{L} \times 0.332 \times k \left(\frac{U}{\nu}\right)^{0.5} (Pr)^{0.333} \times \int_0^L x^{-1} \times x^{0.5} dx$$

$$= \frac{0.332}{L} \times k \times \left(\frac{U}{\nu}\right)^{0.5} \times (Pr)^{0.333} \times \int_0^L x^{-0.5} dx$$

$$= \frac{0.332}{L} \times k \times \left(\frac{U}{\nu}\right)^{0.5} \times (Pr)^{0.333} \times \left[\frac{x^{-0.5+1}}{-0.5+1}\right]_0^L$$

$$= \frac{0.332}{L} \times k \times \left(\frac{U}{\nu}\right)^{0.5} \times (Pr)^{0.333} \times \left[\frac{L^{0.5}}{0.5}\right]$$

$$= \frac{0.332}{0.5} \times \left(\frac{k}{L}\right) \times \left(\frac{UL}{\nu}\right)^{0.5} (Pr)^{0.333}$$

$$= 0.664 \left(\frac{k}{L}\right) (Re)^{0.5} (Pr)^{0.333}$$

$$\text{Average heat transfer} = 0.664 \left(\frac{k}{L}\right) (Re)^{0.5} (Pr)^{0.333} \qquad ....(2.4)$$

Step 3

We know that,

$$\text{Average Nusselt Number, } Nu = \frac{hL}{k}$$

$$\Rightarrow Nu = \frac{0.664 \left(\frac{k}{L}\right) (Re)^{0.5} (Pr)^{0.333} \times L}{k}$$

$$\Rightarrow Nu = 0.664 (Re)^{0.5} (Pr)^{0.333}$$

$$\text{Average Nusselt Number, } Nu = 0.664 (Re)^{0.5} (Pr)^{0.333} \qquad .....(2.5)$$

From equation (2.2) and (2.4), we Know that,

$$h = 2h_x$$

2.10. The Local and Average Heat Transfer Coefficients for Flat Plate – Turbulent Flow

Step 1

The heat transfer coefficient for turbulent flow can be derived using Colburn analogy, From colburn analogy, we know that,

$$St_x(Pr)\, 2/3 = \frac{C_{fx}}{2} = \frac{0.0592}{2}(Re_x)^{-0.2}$$

[From HMT data book, Page No.113]

$$St_x(Pr)\, 2/3 = \frac{0.0592}{2}\times(Re_x)^{-0.2}$$

$$\frac{Nu_x}{Re_x\,Pr}\times[Pr]^{2/3} = 0.0296(Re_x)^{-0.2}$$

$$\frac{Nu_x}{Re_x\,Pr}\times(Pr)^{-1/3} = 0.0296(Re_x)^{-0.2}$$

$$Nu_x(Pr)^{-1/3} = 0.0296\times(Re_x)\,(Re_x)^{-0.2}$$

$$Nu_x = 0.0296\;(Re_x)^{0.8}(Pr)^{1/3}$$

$$\boxed{\text{Local Nusselt No. } (Nu_x)\ = 0.0296(Re)^{0.8}(Pr)^{0.333}}$$

................(2.6)

We know that,

$$Nu_x = \frac{h_x\,x}{k}$$

$$\frac{h_x\,x}{k} = 0.0296\;(Re_x)^{0.8}(Pr)^{0.333}$$

$$h_x = \frac{0.0296(Re_x)^{0.8}\times(Pr)0.333\times k}{x}$$

Local heat transfer coefficient, $h_x =$

$$h_x = 0.0296\left(\frac{k}{x}\right)(Re_x)^{0.8}(Pr)^{0.333}$$

................(2.7)

Step 2

The average heat transfer coefficient, h is given by

$$h = \frac{1}{L} \int_0^L h_x \, d_x$$

$$= \frac{1}{L} \int_0^L 0.0296 \, \frac{k}{x} \, (Re_x)^{0.8} \, (Pr)^{0.333} \, dx \qquad \left[\therefore \ Re = \frac{Ux}{\nu} \right]$$

$$= \frac{0.0296}{L} \times k \times \left(\frac{U}{\nu} \right)^{0.8} (Pr)^{0.333} \int_0^L \left(\frac{1}{x} \right) x \, x^{0.8} \, dx$$

$$= 0.0296x \left(\frac{k}{L} \right) \times \left(\frac{U}{\nu} \right)^{0.8} \times (Pr)^{0.333} \int_0^L x^{-0.2} \, dx$$

$$= 0.0296 \, x \left(\frac{k}{L} \right) \times \left(\frac{U}{\nu} \right)^{0.8} \times (Pr)^{0.333} \left[\frac{x - 0.2 + 1}{-0.2 + 1} \right]_0^L$$

$$= 0.0296 \, x \left(\frac{k}{L} \right) \times \left(\frac{U}{\nu} \right)^{0.8} \times (Pr)^{0.333} \left[\frac{L^{0.8}}{0.8} \right]$$

$$= 0.037 \left(\frac{k}{L} \right) \left(\frac{UL}{\nu} \right)^{0.8} (Pr)^{0.333}$$

$$= 0.037 \left(\frac{k}{L} \right) Re^{0.8} (Pr)^{0.333} \qquad \left[\therefore \ Re = \frac{UL}{\nu} \right]$$

Average heat transfer coefficient, $h_x = 0.037 \left(\frac{k}{L} \right) (Re)^{0.8} (pr)^{0.333}$ .... (2.8)

Step 3

We know that,

Average Nusselt Number, $Nu = \dfrac{hL}{k}$

$$Nu = \frac{0.037 \left(\dfrac{k}{L} \right) (Re)^{0.8} (Pr)^{0.333} \times L}{k}$$

Average Nusselt Number, $Nu = 0.037 \, (Re)^{0.8} \, (Pr)^{0.333}$ .... (2.9)

From equation (2.7) and (2.8), we know that,

Average heat transfer Coefficient, h = 1.25 x h_x

$$h = 1.25 h_x$$

2.10.1. Heat Transfer Coefficient for Combination of Laminar and Turbulent Flow

Step 1

Heat transfer coefficient for laminar –turbulent combined flow is given by

$$h \frac{1}{L} \int_0^x h_x \, dx + \frac{1}{L} \int_x^L h_x \, dx$$

(Laminar) (Turbulent)

$$= \frac{1}{L} \left[\int_0^x 0.332 \left(\frac{k}{x} \right) (Re)^{0.5} (Pr)^{0.333} \, dx + \int_x^L 0.0296 \left(\frac{k}{x} \right) (Re)^{0.8} (Pr)^{0.333} \, dx \right]$$

$$= \frac{1}{L} \left[\int_0^x 0.332 \left(\frac{k}{x} \right) \left(\frac{Ux}{v} \right)^{0.5} (Pr)^{0.333} \, dx + \int_x^L 0.0296 \left(\frac{k}{x} \right) \left(\frac{Ux}{v} \right)^{0.8} (Pr)^{0.333} \, dx \right]$$

$$\left[\therefore Re = \frac{Ux}{v} \right]$$

$$= \frac{k}{L} (Pr)^{0.333} \left[0.332 \left(\frac{U}{v} \right)^{0.5} \int_x^L \frac{x^{0.5}}{v} \, dx + 0.0296 \left(\frac{U}{v} \right)^{0.8} \int_x^L \frac{x^{0.8}}{x} \, dx \right]$$

$$= \frac{k}{L} (Pr)^{0.333} \left[0.332 \left(\frac{U}{v} \right)^{0.5} \int_0^x \frac{x^{0.5}}{v} \, dx + 0.0296 \left(\frac{U}{v} \right)^{0.8} \int_x^L \frac{x^{0.2}}{x} \, dx \right]$$

$$= \frac{k}{L} (Pr)^{0.333} \left[0.332 \left(\frac{U}{v} \right)^{0.5} \left[\frac{x^{-0.5+1}}{-0.5+1} \right]_0^x + 0.0296 \left(\frac{U}{v} \right)^{0.8} \left[\frac{x^{-0.2+1}}{-0.2+1} \right]_x^L \right]$$

$$= \frac{k}{L} (Pr)^{0.333} \left[0.332 \left(\frac{U}{v} \right)^{0.5} \left[\frac{x^{0.5}}{0.5} \right] + 0.0296 \left(\frac{U}{v} \right)^{0.8} \left[\frac{L^{0.8+1}}{0.8} - \frac{x^{0.8}}{0.8} \right] \right]$$

$$= \frac{k}{L} (Pr)^{0.333} \left[\frac{0.332}{0.5} \left(\frac{Ux}{v} \right)^{0.5} + \frac{0.0296}{0.8} \left(\frac{UL}{v} \right)^{0.8} - \frac{0.0296}{0.8} \left(\frac{Ux}{v} \right)^{0.8} \right]$$

$$h = \frac{k}{L} (Pr)^{0.333} \left[0.664 (Re_x)^{0.5} + 0.037 (Re_L)^{0.8} - 0.037 (Re_x)^{0.8} \right]$$

Transition occurs at critical Reynolds number, $Re_0 = 5 \times 10^5$, i.e., Flow is laminar upto Re=5 x105, after that flow is turbulent,

Substitute $Re_0 = Re_x = 5 \times 10^5$

$$\Rightarrow h = \frac{k}{L} \, (Pr)^{0.333} \left[0.664 \, (5 \times 10^5)^{0.5} + 0.037 \, (Re_L)^{0.8} - 0.037 \, (5 \times 10^5)^{0.8} \right]$$

$$= \frac{k}{L} \, (Pr)^{0.333} \left[0.037 \, (Re_L)^{0.8} - 871 \right]$$

> Average heat transfer Coefficient, (h_x)
> $$= h = \frac{k}{L} \, (Pr)^{0.333} \left[0.037 \, (Re_L)^{0.8} - 871 \right]$$

............(2.10)

Step 2

We know that,

Average Nusselt Number, $Nu = \dfrac{hL}{k}$

$$\Rightarrow Nu = \frac{k}{L} \, (Pr)^{0.333} \left[0.037 \, (Re_L)^{0.8} - 871 \right] \times L$$

> Average Nusselt Number, $Nu = (Pr)^{0.333} \left[0.037 \, (Re_L)^{0.8} - 871 \right]$

.... (2.11)

2.11. Boundary Layer Thickness, Shear Stress and Skin Friction Coefficient for Turbulent Flow

Step 1

We know that, Von Karman momentum equation for boundary layer flow is

$$\frac{\tau_0}{\rho U^2} = \frac{d}{dx} \left[\int_0^{\delta} \frac{u}{U} \left(1 - \frac{u}{U} \right) dy \right]$$

Substitute, $\dfrac{u}{U} = \left(\dfrac{y}{\delta} \right)^{1/7}$

$$\Rightarrow \frac{d}{dx} = \left[\int_0^{\delta} \left(\frac{y}{\delta} \right)^{1/7} - \left(\frac{y}{\delta} \right)^{2/7} dy \right]$$

$$= = \frac{d}{dx}\left[\frac{7}{8}\delta \, - \frac{7}{9}\delta \right]$$

$$= \frac{d}{dx}\left\{ \frac{1}{\delta^{1/7}} \int_0^\delta y^{1/7}\,dy - \frac{1}{\delta^{2/7}} \int_0^\delta y^{2/7}\,dy \right\}$$

$$= \frac{d}{dx}\left\{ \frac{1}{\delta^{1/7}} \left(\frac{y^{\frac{1}{7}+1}}{\frac{1}{7}+1} \right)_0^\delta - \frac{1}{\delta^{1/7}} \left(\frac{y^{\frac{2}{7}+1}}{\frac{1}{7}+1} \right)_0^\delta \right\}$$

$$= \frac{d}{dx}\left\{ \frac{1}{\delta^{1/7}} \left(\frac{y^{8/7}}{8/7} \right)_0^\delta - \frac{1}{\delta^{2/7}} \left(\frac{y^{9/7}}{9/7} \right)_0^\delta \right\}$$

$$= \frac{d}{dx}\left[\frac{7}{8}\left(\frac{\delta^{8/7}}{\delta^{1/7}} \right) - \frac{7}{9}\left(\frac{\delta^{9/7}}{\delta^{2/7}} \right)_0^\delta \right]$$

$$= \frac{d}{dx}\left[\frac{7}{8}\delta \, - \frac{7}{9}\delta \right]$$

$$= \frac{d}{dx}\left[\frac{63\delta - 56\delta}{72} \right]$$

$$0.072\,(Re)^{-1/5}$$

$$\boxed{ \frac{\tau_0}{\rho U^2} = \frac{7}{72}\frac{d\delta}{dx} }$$

$$\Rightarrow \tau_0 = \frac{7}{72}\rho U^2 \frac{d\delta}{dx}$$

$$\dots\dots\dots\dots (2.12)$$

Step 2

We know that,

$$\tau_0 = 0.0225\,\rho\,U^2 \left(\frac{\mu}{\rho U \delta} \right)^{1/4} \qquad\qquad \dots\dots\dots\dots (2.13)$$

Equating equation (2.12) and (2.13),

$$\Rightarrow = \frac{0.0225}{(0.370)^{1/4}}\rho U^2 \left[\frac{\nu}{Ux}(Re)^{0.2} \right]^{1/4} \left[\therefore Re = \frac{Ux}{\nu} \right]$$

$$\Rightarrow \frac{7}{72}\frac{d\delta}{dx} = 0.0225\left(\frac{\mu}{\rho U\delta}\right)^{1/4}$$

$$\Rightarrow d\delta = 0.0225\left(\frac{\mu}{\rho U\delta}\right)^{1/4} \times \frac{72}{7}\ dx$$

$$\Rightarrow \delta^{1/4}d\delta = 0.0225\left(\frac{\mu}{\rho U}\right)^{1/4} \times \frac{72}{7}\ dx$$

$$\Rightarrow \delta^{1/4}d\delta = 0.2314\left(\frac{\mu}{\rho U}\right)^{1/4}\ dx$$

Integrating

$$\Rightarrow \int \delta^{1/4}d\delta = 0.2314\left(\frac{\mu}{\rho U}\right)^{1/4} \int dx$$

$$\Rightarrow \frac{\delta^{\frac{1}{4}+1}}{\frac{1}{4}+1} = 0.2314\left(\frac{\mu}{\rho U}\right)^{1/4}\ x + C$$

$$\Rightarrow \frac{\delta^{5/4}}{\frac{5}{4}} = 0.2314\left(\frac{\mu}{\rho U}\right)^{1/4}\ x + C$$

$$\Rightarrow \frac{4}{5}\delta^{5/4} = 0.2314\left(\frac{\mu}{\rho U}\right)^{1/4}\ x + C$$

Assuming boundary layer is turbulent over the entire length the plate.

So, at x=0, δ =0 $\Rightarrow$ C=0

$$\Rightarrow \frac{4}{5}\delta^{5/4} = 0.2314\left(\frac{\mu}{\rho U}\right)^{1/4}\ x$$

$$\Rightarrow \delta^{5/4} = \frac{5}{4} \times 0.2314\left(\frac{\mu}{\rho U}\right)^{1/4}\ x$$

$$\Rightarrow \delta = \left[\frac{5}{4} \times 0.2314\left(\frac{\mu}{\rho U}\right)^{1/4}\ x\right]^{4/5}$$

$$= \left[0.289x\left(\frac{\mu}{\rho U}\right)^{1/4}\ x\right]^{4/5}$$

$$= (0.289)^{4/5} \, x \left(\frac{\mu}{\rho U} \right)^{\frac{1}{4} \times \frac{4}{5}} \, x \; x^{4/5}$$

$$= 0.370 \, X \left(\frac{\mu}{\rho U x} \right)^{1/5} X \; x^{1/5} \; X \; x^{4/5}$$

$$= 0.370 \, X \left(\frac{\mu}{\rho U x} \right)^{1/5} X \; x$$

$$= 0.370 \, X \left(\frac{\nu}{U x} \right)^{1/5} X \; x \qquad \left[\therefore \nu = \frac{\mu}{\rho} \right]$$

$$= 0.370 \, X \left(\frac{1}{Re} \right)^{1/5} X \; x \qquad \left[\therefore Re = \frac{Ux}{\nu} \right]$$

$$\delta = 0.370 \, (Re)^{-0.2} X \; x$$

Boundary layer thickness, $\delta = 0.370 \ (Re)^{-0.2} \times x$

$$\qquad\qquad\qquad\qquad\qquad\qquad\qquad\qquad .... (2.14)$$

Shear stress $\iota_0 = 0.0225 \, \rho U^2 \left(\frac{\mu}{\rho U \delta} \right)^{1/4}$

Substituting δ value,

$$\Rightarrow \tau_0 = 0.0225 \rho U^2 \left(\frac{\mu}{\rho U x \, 0.370 \, (Re)^{-0.2} X \; x} \right)^{1/4}$$

$$= \frac{0.0225}{(0.370)^{1/4}} \rho U^2 \left[\frac{\mu}{\rho U x} \, x (Re)^{0.2} \right]^{1/4}$$

$$= \frac{0.0225}{(0.370)^{1/4}} \, \rho U^2 \left[\; (Re)^{1/5} \, x \, (Re)^{-1} \right]^{1/4}$$

$$= \frac{0.0225}{(0.370)^{1/4}} \, \rho U^2 \left[\frac{(Re)^{0.2}}{Re} \right]^{1/4} \left[\therefore Re = \frac{Ux}{\nu} \right]$$

$$= \frac{0.0225}{(0.370)^{1/4}} \, \rho U^2 \left[(Re)^{1/5} \, x \, (Re)^{-1} \right]^{1/4}$$

$$= 0.02884 \, \rho U^2 \left[\, (Re)^{-4/5} \right]^{1/4}$$

$$=0.02884 \ \rho U^2 \quad \tau_0 = 0.05769 \frac{\rho U^2}{2}(Re)^{-0.2}$$

$$= \frac{0.05769}{2}\rho U^2 \ [Re]^{-0.2}$$

$$\boxed{\text{Sherar stress, } \tau_0 = 0.05769 \frac{\rho U^2}{2}(Re)^{-0.2}} \qquad \ldots\ldots (2.15)$$

Local Skin Friction Coefficient, C_{fx}:

We know that,

$$\text{Shear stress, } \tau_0 = 0.05769 \frac{\rho U^2}{2}(Re)^{-0.2}$$

$$\text{Also, we know } \tau_0 = C_{fx} \ x \ \frac{\rho U^2}{2}$$

Equating both equations,

$$\Rightarrow 0.05769 \ \frac{\rho U^2}{2}(Re)^{-0.2} = C_{fx} \frac{\rho U2}{2}$$

$$C_{fx}=0.05769 \ (Re)^{-0.2}$$

$$\boxed{\text{Local friction coefficient } C_{fx}=0.05769 \ (Re)^{-0.2}} \qquad \ldots\ldots(2.16)$$

Average Friction Coefficient(C_{ft})

$$=\frac{1}{L} \int_0^L 0.05769 \ (Re)^{-0.2} \, dx$$

$$=\frac{1}{L} \int_0^L 0.05769 \ (Re)^{-1/5} \, dx$$

$$=\frac{1}{L} \int_0^L 0.05769 \left(\frac{Ux}{v}\right)^{-1/5} dx$$

$$= \frac{1}{L} \ x \ 0.05769 x \left(\frac{U}{v}\right)^{-1/5} \int_0^L x^{1/5} \, dx$$

$$= \frac{1}{L} \ x \ 0.05769 x \left(\frac{U}{v}\right)^{-1/5} \left[\frac{x^{\frac{-1}{5}+1}}{\frac{-1}{5}+1}\right]_0^L$$

$$= \frac{1}{L} \times 0.05769 \left(\frac{U}{v}\right)^{-1/5} \frac{L^{4/5}}{\dfrac{4}{5}}$$

$$= \frac{5}{4} \times 0.05769 \left(\frac{U}{v}\right)^{-1/5} \times \frac{1}{L} \times 1^{4/5}$$

$$= 0.072 \left(\frac{U}{v}\right)^{-1/5} \times L^{-1} \times L^{4/5}$$

$$= 0.072 \left(\frac{U}{v}\right)^{-1/5} \times L^{4/5}$$

$$= 0.072 \left(\frac{U}{v}\right)^{-1/5}$$

$$= 0.072 \, (Re)^{-1/5}$$

$$0.072 \, (Re)^{-1/5}$$

$$\boxed{\text{Average friction coefficient } C_{fx} = 0.072 \, (Re)^{-0.2}}$$

............(2.17)

Free (OR) Natural Convection

2.12. Free Convection

If the fluid motion is produced due to change in density resulting from temperature gradients the mode of heat transfer is said to be free or natural convection.

Examples

1. The heating of rooms by use of radiators
2. The cooling of transmission lines, electric transforms and rectifiers

The rate of heat transfer is calculated using the general convection equation given below

$Q = hA \, (T_w - T_\infty)$

Where,

Q	–	Heat transfer in w.
A	–	Area in m²
T_w	–	Pipe surface temperature in ºC
T_∞	-	Fluid temperature in ºC.

$G_r.P_r < 10^9$ laminar flow

$G_r.P_r > 10^9$ turbalent flow

2.13. Solved Problem on a Free Convection (Plate)

1. A large vertical plate 4m height s maintained at 606°C and exposed to atmosphere air at 106°C. Calculate the heat transfer if the plate is 10m wide.

Given

Vertical plate

Length (or) Height, l = 4m

Wall temperature, T_w = 606°C

Air temperature, T_∞ = 106°C

Wide, w = 10m

To Find

Heat transfer, (Q)

Solution

Step 1

Velocity is not given

Film temperature, T_f = $\dfrac{T_w + T_\infty}{2} = \dfrac{606 + 106}{2}$ = 356°C

Step 2

Properties of air at 356°C 350°C

[From HMT Data book page no. 34]

ρ = 0.566 kg/m³
v = 55.46 x 10⁻⁶ m²/s
Pr = 0.676
K = 0.04908 w/mk

Step 3

Co-efficient of thermal expansion β = $\dfrac{1}{T_f \text{ in k}}$

(or)

[Refer HMT Databook page no. 30 for air)

$\beta = \dfrac{1}{356 + 273}$

$\beta = 1.58 \times 10^{-3}$ k⁻¹

Step 4

$$\text{Grashof Number, Gr} = \frac{g \times \beta \times L^3 \times \Delta T}{\upsilon^2}$$

[From HMT Databook page no. 135]

$$\Rightarrow \text{Gr} = \frac{9.81 \times 1.58 \times 10^{-3} \times (4)^3 \times (606 - 106)}{(55.46 \times 10^{-6})^2}$$

$\text{Gr} = 1.61 \times 10^{11}$

$\text{GrPr} = 1.61 \times 10^{11} \times 0.676$

$\text{GrPr} = 1.08 \times 10^{11}$

Since $\text{GrPr} > 10^9$, flow is turbulent

Step 5

For turbulent flow,

Nusselt Number $\text{Nu} = 0.10 \, (\text{GrPr})^{0.333}$

[From HMT Databook page no. 136]

$\Rightarrow \text{Nu} = 0.10 \, [1.08 \times 10^{11}]^{0.333}$

$\Rightarrow \text{Nu} = 472.20$

We know that,

$$\text{Nusselt Number, Nu} = \frac{hL}{k}$$

$$\Rightarrow 472.20 = \frac{h \times 4}{0.04908}$$

Heat transfer co-efficient, $h = 5.78 \text{ w/m}^2\text{k}$

Step 6

Heat transfer $Q = hA\Delta T$

$\qquad\qquad = h \times w \times L \times (T_w - T_\infty)$

$\qquad\qquad = 5.78 \times 10 \times 4 \times (606 - 106)$

$\qquad\quad Q = 115600 \text{ w}$

$\qquad\quad Q = 115.6 \times 10^3 \text{ w}$

Result

Heat transfer $Q = 115.6 \times 10^3 \text{ w}$

2. A hot plate 1.2m wide, 0.35m height and at 115°C is exposed to the ambient still air at 25°C. Calculate the following.

(i) Maximum velocity at 180mm from the leading edge of the plate

(ii) The boundary layer thickness at 180mm from the leading edge of the plate.

(iii) Local heat transfer coefficient at 180mm from the leading edge of the plate.

(iv) Average heat transfer co-efficient over the surface of the plate.

(v) Total mass flow through the boundary

(vi) Heat loss from the plate

(vii) Rise in temperature of the air passing through the boundary.

Use approximate solution.

Given

Wide,	$w = 1.2$ m
Height or length,	$L = 0.35$m
Plate surface temperature,	$T_w = 115$°C
Fluid temperature,	$T_\infty = 25$°C
Distance,	$x = 180$mm $= 0.180$m

To Find

i. Maximum velocity at 180mm from the leading edge of the plate, u_{max}.

ii. The boundary layer thickness at 180mm from the leading edge of the plate, δx

iii. Local heat transfer coefficient at 180mm from the leading edge of the plate, h_x.

iv. Average heat transfer co-efficient over the surface of the   plate, h.

v. Total mass flow through the boundary, $\dot{m}$

vi. Heat loss from the plate, Q

vii. Rise in temperature of the air passing through the   boundary, ΔT.

Solution

Step 1

Velocity is not given

$$\text{Fluid temperature, } T_f = \frac{T_w + T_\infty}{2}$$

$$= \frac{115 + 25}{2}$$

$$T_f = 70°C$$

Step 2

Properties at air at 70°C

[From HMT Data book page no. 34]

$\rho = 1.029$ kg/m³

$\upsilon = 20.02 \times 10^{-6}$ m²/s

$Pr = 0.694$

$K = 0.02966$ w/mk

Step 3

Co-efficient of thermal expansion (β)

$$P_{(for\ air)} = \frac{1}{T_f\ ink} = \frac{1}{70+273} = \frac{1}{343}$$

$\beta \quad = 2.91 \times 10^{-3}$ k⁻¹

(or)

[Refer HMT Data book page no – 30]

Step 4

$$\text{Grash of number, Gr} \quad = \frac{g \times \beta \times L^3 \times \Delta T}{\upsilon^2}$$

[From HMT Databook page no. 135]

$$= \frac{9.81 \times 2.91 \times 10^{-3} \times (0.35)^3 \times (115-25)}{(20.02 \times 10^{-6})^2}$$

$Gr_L \quad = 27.5 \times 10^7$

$Gr_L Pr = 27.5 \times 10^7 \times 0.694$

$Gr_L Pr = 1.90 \times 10^8$

Since $Gr_L Pr < 10^9$, for laminar

i.e, $10^4 < Gr_L Pr < 10^9$

Step 5

For laminar flow

Nusselt Number, $Nu = 0.59 (Gr\ Pr)^{0.25}$

[From HMT data book page no. 136]

$Nu = 0.59 (1.90 \times 10^8)^{0.25}$

$Nu = 69.26$

We know that,

Nusslet Number, $Nu = \dfrac{hL}{k}$

$69.26 = \dfrac{hL}{k}$

$69.26 = \dfrac{h \times 0.35}{0.02966}$

$h = 5.86 \ w/m^2k$

Average heat transfer co-efficient over the surface of the plate, $h = 5.86 \ w/m^2k$

Step 6

Heat loss from the plate, Q: $Q = hA \ (T_w - T_\infty)$

For both sides

$Q = 2 \times h \times A \times (T_w - T_\infty)$

$$[\because \ A = W \times L]$$

$Q = 2 \times 5.86 \times (0.35 \times 1.2) \times (115 - 25)$

$Q = 443.01 \ w$

Step 7

For r = 0.180m

Grushaf number $Gr \quad = \dfrac{g \times \beta \times x^3 \times \Delta T}{\upsilon^2}$

$$\text{(From HMT Data book page no. 135)}$$

$= \dfrac{9.81 \times 2.91 \times 10^{-3} \times (0.18)^3 (115 - 25)}{(20.02 \times 10^{-6})^2}$

$Gr = 37.4 \times 10^6$

Step 8

Maximum velocity at 180mm from the leading edge, u_{max}:

$$u_{max} = 0.766 \times v \ (0.952 + Pr)^{-1/2} \times \left[\frac{g\beta \ (T_w - T_\infty)}{\upsilon^2} \right]^{1/2} \times \ x^{1/2}$$

$= 0.766 \times 20.02 \times 10^{-6} \ [0.952 + 0.694]^{-1/2} \times$

$$\left[\frac{9.81 \times 2.91 \times 10^{-3} \times (115 - 25)}{(20.02 \times 10^{-6})^2} \right]^{1/2} \times (0.18)^{1/2} => u_{max} = 0.406 \ m/s.$$

Step 9

The boundary layer thickness at 180mm from the leading edge of the plate, δ_x

$\delta_x = 3.93 \times x \times Pr^{-0.5} (0.952 + Pr)^{0.25} (Gr)^{-0.25}$

[From HMT Databook page no.135]

$= 3.93 \times 0.180 \times (0.694)^{-0.5} \times (0.952 + 0.694)^{0.25} \times$

$(37.4 \times 10^6)^{-0.25}$

$\Rightarrow \delta_x = 0.01229m$

Step 10

Local heat transfer co-efficient at 180mm from the leading edge of the plate, h_x:

$$h_x = \frac{2k}{\delta_x}$$

[From HMT databook, page no. 135]

$$\Rightarrow h_x = \frac{2 \times 29.66 \times 10^{-3}}{0.01229} \Rightarrow h_x = 4.82 \ w/m^2k$$

Step 11

Total mass flow through the boundary ($m^{\bullet}$):

We know that, $m^{\bullet} = 1.7 \times \rho x v \left[\dfrac{Gr_L}{(Pr)^2 \ (Pr + 0.952)} \right]^{0.25}$

$m^{\bullet} = 1.7 \times 1.029 \times 20.02 \times 10^{-6} \times \left[\dfrac{27.5 \times 10^7}{(0.694)^2 (0.694 + 0.952)} \right]^{0.25}$

$m^{\bullet} = 0.00478 \ kg/s$

Step 12

Rise in temperature of the air passing through the boundary (ΔT)

Heat loss, $Q = m \ Cp\Delta T$

$443.01 = 0.00478 \times 1005 \times \Delta T$

$\Rightarrow \Delta T = 92.21k$

Result

i. u_{max} =   0.406 m/s

ii. δ_x =   0.01229 m

iii. h_x =   4.82 w/m²k

iv.	h	=	5.86 w/m²k
v.	$\dot{m}$	=	0.00478 kg/s
vi.	Q	=	443.01w
vii.	ΔT	=	92.21 k

3. A thin 100cm long and 10cm wide horizontal plate is maintained at a uniform temperature of 150°c in a large tankfull of water at 75°C. Estimate the rate of heat to be supplied to the plate to maintain constant plate temperature as heat is dissipated from either side of plate.

Given

Horizontal plate (Water)

Length of horizontal plate, L = 100cm = 01m

Wide, W = 10cm = 0.10m

Plate temperature, T_w = 150°C

Fluid temperature, T_∞ = 75°C

To Find

Heatloss (Q) from either side of plate

Solution

Step 1

Velocity is not given

Fluid temperature, $T_f = \dfrac{T_w + T_\infty}{2} = \dfrac{150 + 75}{2}$

T_f = 112.5°C

Step 2

Properties at air at 112.5°C

[From HMT Data book page no. 34]

ρ = 951 kg/m³

υ = 0.264 x 10⁻⁶ m²/s

Pr = 1.55

K = 0.683 w/mk

Step 3

$\beta_{(for\ water)}$ = 0.8225 x 10⁻³ k⁻¹

[From HMT Data book page No. 30]

Step 4

$$\text{Grash of number, } Gr = \frac{g \times \beta \times L_C^3 \times \Delta T}{\upsilon^2}$$

For horizontal plate characteristics length, $L_c = \dfrac{W}{2} = \dfrac{0.10}{2}$

$L_c = 0.05m$

$$Gr = \frac{9.81 \times 0.8225 \times 10^{-3} \times (0.05)^3 \times (150 - 75)}{(0.264 \times 10^{-6})^2}$$

$Gr_L = 1.0853 \times 10^9$

$Gr_L Pr = 1.0853 \times 10^9 \times 1.55$

$Gr_L Pr = 1.682 \times 10^9$

GrPr value is in between 8×10^6 and 10^{11}

i.e,,$8 \times 10^6 < Gr_L Pr < 10^{11}$

Step 5

For horizontal plate

Nusselt Number, $Nu = 0.15 \, (Gr \, Pr)^{0.333}$

[From HMT Data book page No. 136]

$Nu = 0.15 \, (1.682 \times 10^9)^{0.333}$

$Nu = 177.13$

We know that,

$$\text{Nusslet Number, } Nu = \frac{h_u . Lc}{k}$$

$$177.13 = \frac{h_u \times 0.05}{0.683}$$

$h_u = 2419.7 \, w/m^2 k$

Lower surface heated:

Nusselt Number (Nu)

$Nu = 0.27 \, [GrPr]^{0.25}$

[From HMT Data book page No. 137]

$\Rightarrow \quad Nu = 0.27 \, [1.68 \times 10^9]^{0.25}$

$Nu = 54.68$

We know that

$$Nu = \frac{hlc}{k}$$

$$54.68 = \frac{hl \times 0.05}{0.683}$$

$h_l = 746.94 \ w/m^2k$

Step 6

Total heat transfer, Q

$$Q = (hu + hl) \times A \times \Delta T$$
$$= (hu + hl) \times w \times L \times \Delta T$$
$$= (2419.7 + 746.94) \times 0.10 \times 1 \times (150 - 75)$$
$$Q = 23{,}749.8 \ w$$

Result

Heat transfer, Q = 23,749.8 w

4. Air flow through along rectangular (30cm ht x 60 cm width) air conditioning duct maintains the outer duct surface temperature at 15°C. If the duct is uninsulated and exposed to air at 25°C, calculate the heat gained by the duct per metre length, assuming it to be horizontal.

Given

Air

L = 30cm = 0.3m

W = 60cm = 0.6

$T_w = 15°C$

$T_\infty = 25°C$

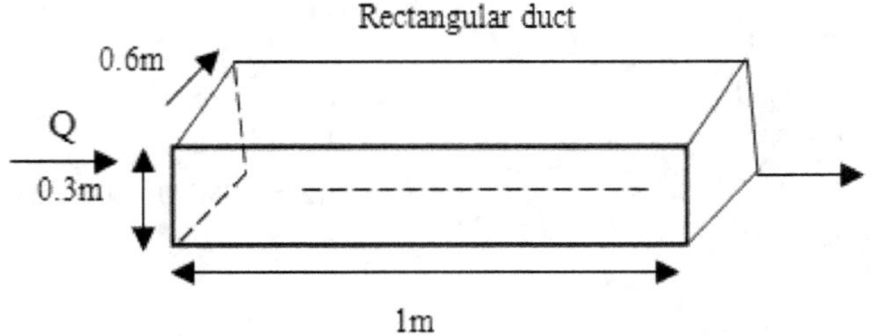

To Find

Heat gained by duct per m length Q/m

Solution

Step 1

Velocity is not given

Film temperature, T_f $= \dfrac{T_w + T_\infty}{2} = \dfrac{15 + 25}{2}$

T_f $= 20$°C

Step 2

Properties at air at 20°C

[From HMT Data book page no. 34]

$\rho = 1.205$ kg/m³

$v = 15.06 \times 10^{-6}$ m²/s

$Pr = 0.703$

$Cp = 1005$ J/kg k

$K = 0.02593$ w/mk

Step 3

$P_{(for\ air)} = \dfrac{1}{T_f\ ink}$

$= \dfrac{1}{20 + 273} = \dfrac{1}{293k}$

β $= 3.42 \times 10^{-3}$ k⁻¹

(or)

[Refer HMT Data book page no – 30]

Step 4

Case (i)

For vertical plate

Grash of number, Gr $= \dfrac{g \times \beta \times L^3 \times \Delta T}{v^2}$

[From HMT Databook page no. 135]

$$= \frac{9.81 \times 3.42 \times 10^{-3} \times (0.3)^3 \times (25-15)}{(15.06 \times 10^{-6})^2}$$

$Gr_L = 3.99 \times 10^7$

$GrPr = 3.99 \times 10^7 \times 0.703$

$Gr\,Pr = 2.80 \times 10^7$

Since $Gr_L Pr < 10^9$, for laminar flow

i.e,, $10^4 < Gr_L Pr < 10^9$

Step 5

For laminar flow

Nusselt Number, $Nu = 0.59\,(Gr\,Pr)^{0.25}$

[From HMT data book page no. 136]

$Nu = 0.59\,(2.80 \times 10^7)^{0.25}$

$Nu = 43$

We know that,

Nusslet Number, $Nu = \dfrac{h_v L}{k}$

$$h_v = \frac{43 \times 0.3}{0.02593}$$

$h = 3.71\ w/m^2 k$

Step 6

Case (ii) Horizontal Plate

Grash of number, $Gr = \dfrac{g \times \beta \times L_c^{\,3} \times \Delta T}{v^2}$

[From HMT Databook page no. 135]

Characteristic length $Lc = W/2$

$$Lc = \frac{0.6}{2} = 0.3$$

$$= \frac{9.81 \times 3.42 \times 10^{-3} \times (0.3)^3 \times (25-15)}{(15.06 \times 10^{-6})^2}$$

$Gr_L = 3.98 \times 10^7$

$Gr_L Pr = 3.98 \times 10^7 \times 0.703$

$Gr_L Pr = 2.79 \times 10^7$

Step 7

For Horizontal Plate

Upper Surface

Nusslet number,

$Nu = 0.15 (GrPr)^{0.25}$

[From HMT data book page no. 136]

$Nu = 0.15 (2.79 \times 10^7)^{0.25}$

$Nu = 45.28$

We know that

$$NussletNumber\ Nu = \frac{h_u L_c}{k}$$

$$h_u = \frac{45.28 \times 0.2593}{0.3}$$

$h_u = 3.89\ w/m^2 k$

Lower Surface

Nusselt Number

$Nu = 0.27 (GrPr)^{0.25}$

[From HMT Databook page no. 137]

$Nu = 0.27 (2.79 \times 10^7)^{0.25}$

$Nu = 19.62$

We know that

$$Nusselt\ Number\ Nu = \frac{h_L . L_c}{k}$$

$$hl = \frac{19.62 \times 0.02593}{0.3}$$

$hl = 1.70\ w/m^2 k$

Step 8

Total Heat Transfer

$Q = (2 \times h_v\ A\ \Delta T) + [(h_u + h_l)\ A\ \Delta T]$

$Q = [2 \times h_v \times w \times L \times (T_w - T_\infty)] + [(h_u + h_l) \times w \times L \times (T_w - T_\infty)]$

$Q = (2 \times 3.7 \times 0.3 \times 1 \times 10) + [(3.92 + 1.71) \times 0.6 \times 1 \times 1 \times 10)]$

$Q = 55.92\ w/m$

2.14. Solved Problems on a Free Convection (Cylinder)

1. A steam pipe 10 cm outside diameter runs horizontally in a room at 23°C. Take the outside surface temperature of pipe as 165°C. Determine the heatloss per metre length of pipe.

Given

Diameter of the pipe, $D = 10cm = 0.10m$

Ambient air temperature, $T_\infty = 23°C$

Wall temperature, $T_w = 165°C$

To Find

Heat loss per metre length

Solution

Step 1

Velocity is not given

Film temperature, $T_f = \dfrac{T_w + T_\infty}{2} = \dfrac{165 + 23}{2}$

$$T_f = 94°C$$

Step 2

Propertise of air at $94°C \approx 95°C$

[From HMT Data book page no. 34]

$\rho = 0.959 \text{ kg/m}^3$

$v = 22.615 \times 10^{-6} \text{ m}^2/\text{s}$

$Pr = 0.689$

$K = 0.03169 \text{w/mk}$

Step 3

Co-efficient of thermal expansion β

$$\beta = \frac{1}{T_f \text{ in k}}$$

$$\beta = \frac{1}{94 + 273}$$

$\beta = 2.72 \times 10^{-3} \text{ k}^{-1}$

Step 4

$$\text{Grashof Number, } Gr = \frac{g \times \beta \times D^3 \times \Delta T}{v^2}$$

[From HMT Databook page no. 135]

$$\Rightarrow Gr = \frac{9.81 \times 2.72 \times 10^{-3} \times (0.10)^3 \times (165-23)}{(22.615 \times 10^{-6})^2}$$

Gr	$= 7.40 \times 10^6$
GrPr	$= 7.40 \times 10^6 \times 0.689$
GrPr	$= 5.09 \times 10^6$

Step 5

For horizontal cylinder,

Nusselt Number $Nu = C (GrPr)^m$

[From HMT Databook page no. 137]

$\Rightarrow Nu = 0.48 [5.09 \times 10^6]^{0.25}$

Grpr $= 5.09 \times 10^6$, Corresponding C = 0.48, m = 0.25

$\Rightarrow Nu = 22.79$

We know that,

$$\text{Nusselt Number, } Nu = \frac{hD}{k}$$

$$\Rightarrow 22.79 = \frac{h \times 0.10}{0.03169}$$

$h = 7.22 \ w/m^2k$

Step 6

Heat less Q $= hA\Delta T$

$\qquad = h \times \pi \times D \times L \ (T_w - T_\infty)$

Q/L $= 7.22 \times \pi \times 0.10 \times (165 - 23)$

Q/L $= 322.08 \ w/m$

Result

Heat transfer Q/L = 322.08 w/m

2. A vertical pipe of 12cm outer diameter, 2.5m long, at a surface temperature of 120°C is in a room where the air is at 20°C. Calculate the heat loss per meter length of the pipe.

Given

Diameter of the pipe, D = 12cm = 0.12m

Length, L = 2.5m

Room temperature, T_w = 20°C

Surface temperature, $T\infty$ = 20°C

To Find

Heat loss (Q) per metre length of pipe.

Solution

Step 1

Velocity is not given

$$\text{Film temperature, } T_f = \frac{T_w + T_\infty}{2}$$

$$= \frac{120 + 20}{2}$$

$$T_f = 70°C$$

Step 2

Properties of air at 70°C

[From HMT Data book page no. 34]

ρ = 1.029 kg/m³

v = 20.02 x 10⁻⁶ m²/s

P_r = 0.694

K = 0.02966 w/mk

Step 3

$$\beta = \frac{1}{T_f \text{ in } k}$$

$$\beta = \frac{1}{70 + 273}$$

$$\beta = 2.91 \times 10^{-3} \, k^{-1}$$

Step 4

$$\text{Grashof Number, Gr} = \frac{g \times \beta \times D^3 \times \Delta T}{v^2}$$

[From HMT Databook page no. 135]

$$\Rightarrow \text{Gr} = \frac{9.81 \times 2.91 \times 10^{-3} \times (2.5)^3 \times (120-20)}{(20.02 \times 10^{-6})^2}$$

Gr $\qquad = 1.11 \times 10^{11}$

GrPr $\qquad = 1.11 \times 10^{11} \times 0.694$

GrPr $\qquad = 7.72 \times 10^{10}$

Since GrPr $> 10^9$, flow is turbulent.

Step 5

For turbulent flow,

$\text{Nu} = 0.10 \ (\text{GrPr})^{0.333}$

[From HMT Databook page no. 136]

$\Rightarrow \text{Nu} = 0.10 \ [7.72 \times 10^{10}]^{0.333}$

$\qquad \text{Nu} = 422.3$

We know that,

$$\text{Nusselt Number, Nu} = \frac{hL}{k}$$

$$\Rightarrow 422.3 = \frac{h \times 2.5}{0.02966}$$

$h = 5.01 \ \text{w/m}^2\text{k}$

Step 6

Heat loss per metre length $Q = hA\Delta T$

$\qquad = h \times \pi \times D \times L \ (T_w - T_\infty)$

$\qquad = 5.01 \times \pi \times 0.12 \times (120 - 20)$

$Q \qquad = 188.8 \ \text{w/m}$

Result

Heat loss per metre length of pipe, $Q = 188.8 \ \text{w/m}$

2.15. Solved Problems on a Sphere

1. A Sphere of diameter 20mm is at 300°C is immersed in air at 25°C. Calculate the convective heat loss.

Given

Diameter of Sphere, D = 20mm = 0.020m

Surface temperature, T_w = 300°C

Fluid temperature, T_∞= 25°C

To Find

Convective heat loss (Q).

Solution

Step 1

Velocity is not given

Film temperature, $T_f = \dfrac{T_w + T_\infty}{2}$

$$= \dfrac{300 + 25}{2}$$

T_f = 162.5°C

Step 2

Properties of air at 162.5 °C ≈ 160°C

ρ = 0.815 kg/m³

v = 30.09 x 10⁻⁶ m²/s

Pr = 0.682

K = 0.03640 w/mk

Step 3

We know

$$\beta = \dfrac{1}{T_f \text{ in } k}$$

$$\beta = \dfrac{1}{162.5 + 273} = \dfrac{1}{435.5}$$

β = 2.29 x 10⁻³ k⁻¹

Step 4

$$\text{Grashof Number, Gr} = \frac{g \times \beta \times D^3 \times \Delta T}{v^2}$$

[From HMT Databook page no. 135]

$$\Rightarrow \text{Gr} = \frac{9.81 \times 2.29 \times 10^{-3} \times (0.020)^3 \times (300 - 25)}{(30.09 \times 10^{-6})^2}$$

Gr = 54734.2

GrPr = 54734.2 x 0.682

GrPr = 37328.7 $\qquad$ [1 <GrPr< 10^5]

Step 5

For Sphere,

Nusselt Number, Nu = 2 + 0.43 $(GrPr)^{0.25}$

[From HMT Databook page no. 138]

=> Nu = 2 + 0.43 $[37328.7]^{0.25}$

Nu = 7.97

We know that,

$$\text{Nusselt Number, Nu} = \frac{hD}{k}$$

$$\Rightarrow 7.97 = \frac{h \times 0.020}{0.03640}$$

h   =   14.51 w/m²k

Step 6

Heat transfer Q = hAΔT

$$= h \times 4\pi r^2 \times (T_w - T_\infty)$$

$$= h \times 4 \times \pi \times \left(\frac{0.020}{2}\right)^2 \times (300\text{-}25)$$

Q   = 5.01 w

FLOW OVER BANK OF TUBES			
1	To find flim temperature	$T_f = \dfrac{Tw + T\infty}{2}$	112
2	To find properties	From T_f	
		For water	22
		For gaser	34
		To take following properties	
		Density (P) in kg/m^3	
		Kinematic viscosity (V) in m^2/s	
		Prantel number (Pr)	
		Thermal conductivity (K) in W/mk	
3	To find maximum velocity (U_{max})(m/s)	$U_{max} = U * \dfrac{St}{St - D}$	
		St = transverse pitch ,m	
		U = velocity of the fluid , m/s	
		D = diameter of tube , m	
4	To find Reynolds number	$Re = (U_{max}.D)/V$	
5	To find nusselt number	Nu = 1.13 * (Pr)$^{0.33}$ * [CRe^n]	
		To find C and n values derived	123
		From St/D and Sl/D ratio respectively	
		Sl-longitudinal pitch , m	

1. When 0.5 kg of water per minute is passed through a tube of 2.5 cm diameter, it is found to be heated from 20°C to 40°C. The heating is achieved by condensing steam on the surface of the tube and subsequently the surface temperature of the pipe is maintained at 110°C. Determine the length of the tube required for fully developed flow.

Given

Inlet temperature, T_{mi} = 20°C

Outlet temperature, T_{mo} = 40°C

Diameter, D = 2.5cm = 0.025m

Pipe surface temperature, T_w = 110°C

Mass flow rate, $m^\bullet$ = 0.5 kg/min = 8.33 x 10^{-3} kg/s

To Find

Length of the tube (L)

Solution

Step 1

Bulk mean temperature, T_m $= \dfrac{T_{mi} + T_{m0}}{2} = \dfrac{20 + 40}{2}$

$$T_m = 30°C$$

Step 2

Properties of water at 30°C

[From HMT Databook page no. 22]

$\rho = 997 \ kg/m^3$

$v = 0.857 \times 10^{-6} \ m^2/s$

$Pr = 5.5$

$K = 0.610 \ w/mk$

$Cp = 4178 \ J/kg \ k$

Step 3

Mass flour rate, $m^{\bullet} = \rho AU$

$$8.33 \times 10^{-3} = \rho \times \frac{\pi}{4} D^2 \times U$$

$$8.33 \times 10^{-3} = 997 \times \frac{\pi}{4} \times (0.025)^2 \times U$$

$U = 0.017 \ m/s$

Reynold number $= \dfrac{UD}{v} = \dfrac{0.017 \times 0.025}{0.857 \times 10^{-6}}$

$Re = 495$

Since Re < 2300, flow is laminar

Step 4

For Laminar flow,

Nusselt Number Nu = 3.66

[HMT Data book page no. 124]

We know that

$$Nu = \frac{hD}{k}$$

$$3.66 = \frac{h \times 0.025}{0.610}$$

$h = 89.3 \ w/m^2k$

Step 5

Heat transfer, $Q = \dot{m}C_p\Delta T$

$$= \dot{m}C_p(T_{mo} - T_{mi})$$

$$= 8.33 \ \times 10^{-3} \times 4178 \ (40 - 20)$$

$$Q = 696.05w$$

Heat transfer $Q = hA \ (T_w - T_m)$

$$= h \times \pi \times D \times L \ (T_w - T_m)$$

$$= 89.3 \times \pi \times 0.025 \times L \ (110 - 30)$$

$$L = 1.24m$$

Result

Length of the tube, $L = 1.24m$

2. In a long annulus (3.125 cm ID and 5cm OD) the air is heated by maintaining the temperature of the outer surface of inner tube at 50°C. The air enters at 16°C and leaves at 32°C. Its flow rate is 30 m/s estimate the heat transfer co-efficient between air and the inner tube.

Given

Inner diameter, D_i = 3.125cm	= 0.03125m
Outer diameter, D_0 = 5cm	= 0.05m
Tube wall temperature, T_w	= 50°C
Inner temperature of air, T_{mi}	= 16°C
Outer temperature of axis, T_{mo}	= 32°C

To Find

Heat transfer co-efficient (h)

Solution

Step 1

Mean temperature, $T_m = \dfrac{T_{mi} + T_{m0}}{2}$

$$= \dfrac{16 + 32}{2}$$

$$T_m \quad = 24°C$$

Step 2

Properties of air at 24°C

$\rho = 1.185 \ kg/m^3$

$v = 15.53 \times 10^{-6} \ m^2/s$

$Pr = 0.702$

$K = 0.02634 \ w/mk$

Step 3

Hydraulic or Equivalent diameter

$$D_h \ = \frac{4A}{p} = \frac{4x\frac{\pi}{4}\left[D_0^{\,2} - D_1^{\,2}\right]}{\pi\left[D_0 + D_i\right]}$$

$$= \frac{\left[D_o + D_i\right]\left[D_o - D_i\right]}{\left[D_o + D_i\right]}$$

$$= D_o - D_i$$

$$= 0.05 - 0.03125$$

$$D_h \ = 0.01875m$$

$$\text{Reynolds number, Re} = \frac{U\,Dn}{v}$$

$$= \frac{30 \times 0.01875}{15.53 \times 10^{-6}}$$

$$Re \quad = 36.2 \times 10^3$$

Since Re > 2300, flow is turbulent

Step 4

For turbulent flow, general equation

(R > 10000)

$Nu = 0.023 \ (Re)^{0.8} \ (Pr)^n$

[From HMT Databook page no. 126]

This is heatingprocess, so n = 0.4

$[\ \therefore T_{mo} > T_{mi}]$

$Nu = 0.023 \times (36.2 \times 10^3)^{0.8} \ (0.702)^{0.4}$

$Nu = 88.59$

We know that

$$\text{Nu} = \frac{hD_h}{k}$$

$$88.59 = \frac{h \times 0.01875}{26.34 \times 10^{-3}}$$

h = 124.4 w/m²k

Result

Heat transfer co-efficient, h = 124.4 w/m²k

3. Engine oil flows through a 50mm diameter tube an average temperature of 147°C. The flow velocity is 80 cm/s. calculate the average heat transfer co-efficient if the tube wall is maintained at a temperature of 200°C and it is 2m long.

Given

Diameter, D = 50mm   = 0.050m

Average temperature, T_m = 147°C

Velocity, U =80 cm/s   = 0.80 m/s

Tube wall temperature, T_w = 200°C

Length, L   = 2m

To Find

Average heat transfer co-efficient (h).

Solution

Step 1

Properties of engine oil at 147°C

[From HMT data book page no. 25]

ρ = 816 kg/m³

$v = 8 \times 10^{-6}$ m²/s

Pr = 116

K = 0.1338 w/mk

Step 2

We know that,

$$\text{Reynolds Number Re} = \frac{UD}{v}$$

$$= \frac{0.8 \times 0.05}{8 \times 10^{-6}}$$

$$\text{Re} = 5000$$

Since Re > 2300 flow is turbulent

$$\frac{L}{D} = \frac{2}{0.050} = 40$$

$$10 < \frac{L}{D} < 400$$

Step 3

For turbulent flow, (Re < 10000)

$$\text{Nusselt Number, } Nu = 0.036 \, (Re)^{0.8} \, (Pr)^{0.33} \left(\frac{D}{L}\right)^{0.055}$$

[From HMT Databook page no. 126]

$$Nu = 0.036 \, (5000)^{0.8} \times (116)^{0.33} \times \left(\frac{D}{2}\right)^{0.055}$$

$$Nu = 128.42$$

We know that,

$$Nu = \frac{hD}{k}$$

$$128.42 = \frac{h \times 0.050}{0.1338}$$

$$h = 343.65 \ w/m^2 k$$

Result

Heat transfer co-efficient, $h = 343.65 \ w/m^2 k$.

Unit 3

PHASE CHANGE HEAT TRANSFER AND HEAT EXCHANGERS

3.1. Boiling

The change of phase from liquid to vapour state is known as boiling.

3.2. Condensation

The change of phase from vapour to liquid state is known as condensation.

3.3. Modes of Condensation

1. Film wise condensation
2. Drop wise condensation

3.3.1. Film Wise Condensation

The liquid condensate wets the solid surface, spreads out and forms a continuous film over the entire surface. Film condensation occurs when a vapour free from impurities.

3.3.2. Dropwise Condensation

In dropwise condensation, the vapour condenses into small liquid droplets of various sizes which fall down the surface in a random fashion. Heat transfer rate may be as much 10 times higher than in film wise condensation.

3.4. Nusselt's Theory for Film Condensation

1. The plate is maintained at a uniform temperature T_w, which is less than the saturation temperature T_{sat} of the vapour.
1. Fluid properties are constant
2. The shear stress at a liquid vapour interface is negligible.
3. The condensing vapour is entirely clean and free from gases, air and non-condensing impurities.
4. The heat transfer across the condensate layer is by pure conduction and the temperature distribution is linear.

3.5. Boiling Heat Transfer Phenomena

Boiling is a convection process involving a change of phase from liquid to vapour state. This is possible only when the temperature of the surface (T_w) exceed's the saturation temperature of liquid (T_{sat}).

According to convection law,

$$Q = hA\,(T_w - T_{sat})$$

(or)

$$Q = hA\,(\Delta T)$$

where,

$\Delta T = (T_w - T_{sat})$ is known as excess temperature.

If heat is added to a liquid from a submerged solid surface, the boiling process is refered to as pool boiling. In this case the liquid above the hot surface is essentrially stagnant and its motion near the surface is due to free convection and mixing induced by bubble growth detachment.

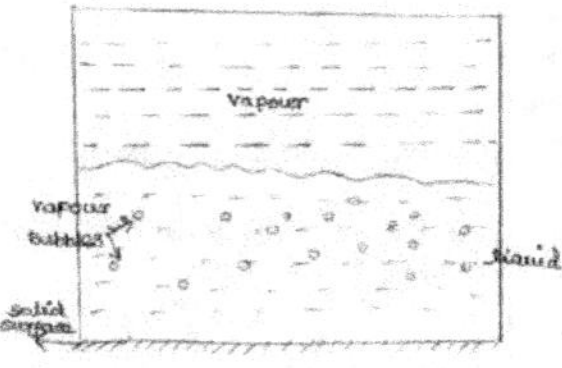

Fig. 3.1: Pool Boiling

The temperature distribution in saturated pool boiling with a liquid – vapour interface.

The different regions of boiling are indicated fig.3.2.

The specific curve has been obtained from an electrically heated platinum wire submerged in a pool of water by varying its surface temperature and measuring the surface heat flux (9).

Pool Boiling Curve for Water

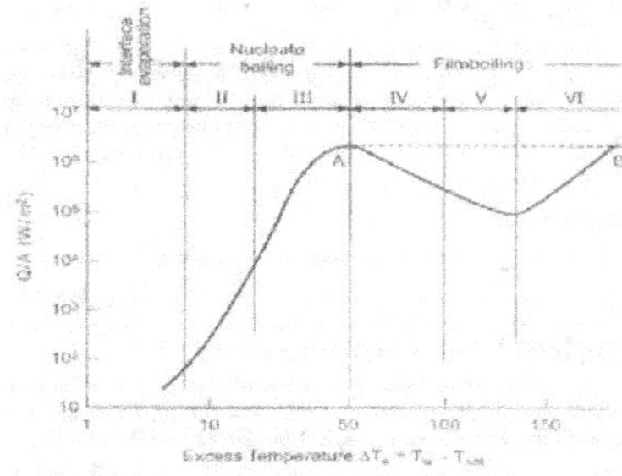

Fig. 3.2: Pool Boiling Curve for Water

I   – Free convection

II   – Bubbles condense in super heated liquid.

III   – Bubbles raise to surface

IV   – Unstable film

V   – Stable film

VI   – Radiation coming into play.

1. Interface Evaporation

- Interface evaporation process with no bubble formation exists in region 1.
- It is have ΔT very low ($5^{\circ}C$)
- Here the liquid near the surface is super heated slightly and evaporation takes place at the liquid surface.

2. Nucleate Boiling

- Boiling exists in region (II) and (III). The nucleate boiling at region II. As the excess temperature to further increased.
- Bubbles are formed rapidly and rapid evaporation takes place.
- Nuclear boiling exists upto $\Delta T = 50^{\circ}C$
- The maximum heat flux, known as critical heat flux occurs at point A. (Fig.3.2)

3. Film Boiling

- In region IV the vapour film is not stable and collapse and reforms rapidly with further increase in ΔT (excess temperature) the vapour film is stabilised as indicated in region V.
- The surface temperature required to maintain a stable film are high and under these conditions a sizeable amount of heat is lost by the surface due to radiation.

3.6. Flow Boiling

- Flow boiling (or) forced convection boiling may occur when a fluid is forced through a pipe or over a surface which is maintained at a temperature higher than the saturation temperature of the fluid.

Example

Water tube boilers involving the forced convection

3.7. Formula (or) Correlations Used in the Boiling and Condensation

Sl.No	Nucleate pool soiling $\Delta T < 50\rho°C$	HMT DB Pg.No	S. no	Film Boiling $\Delta T < 50\rho°C$	HMT DB Pg.No
1	Heat flux $\left(\dot{q} = \frac{Q}{A}\right)$ in $\frac{W}{m^2}$ $$\frac{Q}{A} = \mu_l h_{fg}\left[g * \frac{\rho_l - \rho_v}{\sigma}\right]^{0.5} * \left[\frac{C_{pl} * \Delta T}{C_{sf} * h_{fg}\, p_r^n}\right]^{3.0}$$ Notes:	143	1	Heat transfer coefficient (h) $h = h_{conv} + h_{rad}(0.75)$	143
a)	surface temperature tw is given saturation temperature of water is 100ºC properties of water at 100ºC density ,ρ_l=961 Kg/m³ kinematic viscosity ,V =0.293 * 10⁻⁶m²/s		2	heat transfer co efficient of convective $$h_{conv} = 0.62\left[\frac{k_v^3 \rho_v (\rho_l - \rho_v).g\left(h_{fg} + 0.68 C_{pv}\Delta T\right)}{\mu_v D \Delta T}\right]^{0.25}$$	
b)	prantel number ,Pr = 1.740 specific heat ,C_{pl} = 4216J/kgk Dynamic viscosity = μ_l =ρ_l*V μ_l=281.57*10-6N-S/m² From steam table at 100ºC Enthalpy of evaporation h_{fg} = 2256.9 KJ/Kg h_{fg}=2256.9*10³J/Kg	22	3	Heat transfer coefficient of radiation $$h_{rad} = \sigma_r \varepsilon \left[\frac{Tw^4 - Tsat^4}{Tw - Tsat}\right]$$ Notes	
c)	Specific volume of vapour , Vg = 1.673m³/kg Density of vapour , $\rho_v = \frac{1}{V_g} = \frac{1}{1.673}$ $\rho_v = 0.597 \frac{Kg}{m^3}$ ΔT(Excess temperature) $\Delta T = T_w - T_{sat}$ (ºC)	04	a)	Surface temperature is given (T_w) Saturation temperature of water is 100ºC(J_{sat})	
d)	There fore ΔT < 50ºC ,So nucleate pool boiling		b)	Excess temperature $\Delta T = J_w - T_{sat}$ Therefore ΔT>50ºC, So Film boiling	
e)	σ – Surface tension for liquid vapour interface at 100ºC $\sigma = 0.0588 \frac{N}{m}$		c)	Film temperature $J_f = \frac{T_w + T_{sat}}{2}$ (ºC)	40
f)	Csf –surface fluid constant (Ex : water copper) Csf=0.013	145	d)	Properties of water rapour at Tf(ºC) Density ρ_v in kg/m³ Absolute Viscosity μ_v in $\frac{N-S}{m^2}$ Specific heat C_{pv} in J/kgk Thermal conductivity K_v in W/mK	
g)	n=1 for water n=1.7 for other fluid	144	e)	Properties of water at 100ºC P_l (Density) in kg/m³	22
2	critical heat flux (or) maximum heat flux $\frac{Q}{A}$ in $\frac{w}{m^2}$ $$\frac{Q}{A} = 0.18\, h_{fg} . Pr\left[\frac{\sigma . g . \rho_l - \rho_v}{\rho_v^2}\right]^{0.25}$$ Rate of evaporation (ṁ) in kg/s Heat transfer Q=ṁ*h_{fg} $\dot{m} = \frac{Q}{h_{fg}}$	143	f)	From steam fable at 100ºC h_{fg} = 2256.9KJ/kg h_{fg} = 2256.9*10³ J/kg	
			4	Heat Transferred (Q) in W $Q = hA(T_w - T_{sat})$ $A = \pi QL(m^2)$	
			5	Rate of evaporation (m) in kg/s $\dot{m} = \frac{Q}{hfg}$	

3.8.   Problems Based on the Boiling Type

An Aluminium pan of 15cm diameter is used to boil water and the water depth at a time of boiling is 2.5cm. The pan is placed on an electric stove and the heating element raises the temperature of the pan to 110°C. Calculate the power input for boiling and the rate of evaporating take $C_{sf} = 0.0132$.

Step 1

Given

Diameter, d = 15cm = 0.15m

Distance, x = 2.5cm = 0.025m

Surface of temperature, Tw = 110°C

C_{sf}= 0.0132

Step 2

To Find

1.   Power input (P)
2.   Rate of evaporation (in)

Step 3

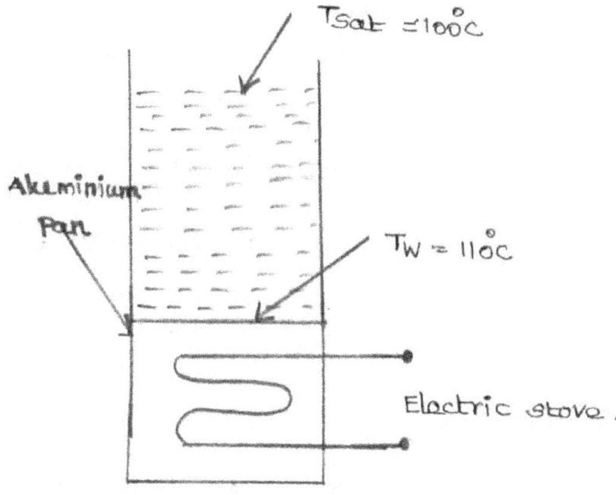

Fig. 3.3

Step 4

We know that,

$$\boxed{\text{Saturation temperature of water } 100^\circ\text{C}}$$

i.e., ($T_{sat} = 100^\circ$C)

Properties of water at 100°C,

{HMTDB P.No 21}

Density, $\rho_l = 961$ kg/m^3

Kinematic viscosity, $v = 0.293 \times 10^{-6}$ m^2/s.

Prandal number, $P_r = 1.740$

Specific heat, $C_{pl} = 4216$ J/kg.K

Dynamic viscosity, $\mu_l = \rho_l \times v = 961 \times 0.293 \times 10^{-6}$

$$\mu_l = 281.57 \times 10^{-6} \text{ N-s/m}^2$$

Step 4

From steam table at 100°C,   (R.S. Khumi steam table P.No.4)

Enthalphyof evaporation,

$h_{fg} = 2256.9$ kJ/kg

$$\boxed{h_{fg} = 2256.9 \times 10^3 \text{ J/kg}}$$

Specific volume of vapour,

$v_g = 1.673$ m^3/kg

Density of vapour ρ_v $= 1/vg = 1/0.673$

                 ρ_v $= 0.597$ kg/m^3

$$\boxed{\Delta T = \text{Excess temperature} = T_w - T_{sat}}$$

$$= 110^\circ\text{C} - 100^\circ\text{C}$$

$\Delta T = 10^\circ$C {$\Delta T = 10^\circ$C $< 50^\circ$C, so this nucleate pool boiling process}

Step 5

For nucleate pool boiling

i. Power input for boiling:

$$\boxed{\text{Heat flux, } \frac{Q}{A} = \mu_1 \times h_{fg} \times \left[\frac{g(\rho_l - \rho_v)}{\sigma}\right]^{0.5} \times \left[\frac{c_{pl} \times \Delta T}{c_{sf} \times h_{fg} \, p_r^n}\right]^{0.3}}$$

where,

n = 1 for water

σ = surface tension for liquid vapour interface.

At 100°C,

σ = 0.0588 N/m

Substitute,

μ_1, h_{fg}, ρ_l, ρ_v, σ, c_{pl}, ΔT, c_{sf}, h_{fg}, n and p_r values in eqn (1)

$$(1)\quad \frac{Q}{A} = 281.57 \times 10\text{-}6 \times (2256.9 \times 10^{+3}) \times \left[\frac{9.81 \times (961 - 0.597)}{0.0588}\right]^{0.5} \times \left[\frac{4216 \times 10}{0.013 \times 2256.9 \times 10^3 \times 1.740}\right]^3$$

Heat transfer, $Q = 1.43 \times 10^5 \times A$

$$Q = 1.43 \times 10^5 \times \pi/4\ (d)^2$$

$$\boxed{\frac{Q}{A} = 1.43 \times 10^5 \text{ w/m}^2}$$

$$Q = 1.43 \times 10^5 \times \pi/4\ (0.15)^2$$

$$\boxed{Q = 2527w = P}$$

Power input for boiling = 2527 watts

Step 6

Rate of Evaporation : $(\dot{m})$

We know that,

$$\boxed{\text{Heat transferred } Q = \dot{m} \times hfg}$$

$$\dot{m} = \frac{Q}{h_{fg}} = \frac{2257}{2256.9 \times 10^3}$$

$$\boxed{\dot{m} = 1.11 \times 10^{-3} \text{ kg/S.}}$$

Step 7

Result

1. $\rho = 2527w$

2. $\dot{m} = 1.11 \times 10^{-3}$ kg/s

1. It is desired to boil water at atmospheric pressure on a copper surface which is electrically heated. Estimate the heat flux from the surface to the water. If the surface is maintained at 110°C and also the peak heat flux.

Step 1

Given Data

Surface temperature, $T_w = 110°C$

Step 2

To find:

i) Heat flex (Q/A)

ii) Critical heat flux (Q'/A)

Step 3

We know that, saturation temperature of water at 100°C.

i.e., $(T_{sat} = 100°C)$

Properties of water at 100°C {HMTDB P.No.21}

Density (ρ_l) $= 961$ kg/m³

Kinematic viscosity, $v = 0.293 \times 10^{-6}$ m²/S.

Prandal number, P_r $= 1.740$

Specific heat, C_{pl} $= 4216$ J/kg.k

Dynamic viscosity μ_l $= \rho_l \times v = 961 \times 0.293 \times 10^{-6}$

$$\mu_l = 281.57 \times 10^{-6} \text{ N-S / m}^2.$$

Step 4

From steam table At 100°C, $\hspace{2cm}$ (R.S. Khumi steam table P.No.4)

* Enthalphy of evaporation ,

$h_{fg} = 2256.9$ kJ/kg

$$h_{fg} = 2256.9 \times 10^3 \text{ J/kg}$$

* Specific volume of vapour,

$$v_g = 1.673 \text{ m}^3/\text{kg}$$

* Density of vapour ρ_v $= 1/v_g$ $= 1/1.673$

$$\rho_v \hspace{1cm} = 0.597 \text{ kg/m}^3$$

* ΔT = Excess temperature $\quad = T_w - T_{sat}$

$$= 110^{\circ}C - 100^{\circ}C$$

$$\boxed{\Delta T = 10^{\circ}C}$$

{$\Delta T = 10^{\circ}C < 50^{\circ}C$, so this nucleate pool boiling process}

Step 5

For nucleate pool boiling

$$\boxed{\text{Heat flux, } \frac{Q}{A} = \mu_l \times h_{fg} \times \left[\frac{g(P_1 - \rho_v)}{\sigma} \right]^{0.5} \times \left[\frac{c_{pl} \times \Delta T}{c_{sf} \times h_{fg}\, p_r^n} \right]^{0.3}}$$

{from HMTDB P.No -1 43}

where,

n = 1 for water

σ = surface tension for liquid vapour interface.

At $100^{\circ}C$,

$$\boxed{\sigma = 0.0588 \text{ N/m}}$$

{HMT DB P.No 45}

$$\boxed{\text{for water - copper} \Rightarrow c_{sf} = \text{surface fluid constant } = 0.013}$$

Substitute,

μ_l, h_{fg}, ρ_l, ρ_v, σ, c_{pl}, ΔT, c_{sf}, h_{fg}, n and p_r values in eqn (1)

(1) $\quad \dfrac{Q}{A} = 281.57 \times 10^{-6} \times (2256.9 \times 10^3) \times \left[\dfrac{9.81 \times (961 - 0.597)}{0.0588} \right]^{0.5} \times \left[\dfrac{4216 \times 10}{0.013 \times 2256.9 \times 10^3 \times 1.740} \right]^3$

$$\boxed{\frac{Q}{A} = 142.83 \times 10^3 \text{ w/m}^2}$$

Step 6

Nucleate Pool boiling

Critical heat flux,

{HMTDB P.No.143}

$$\boxed{\left(\frac{Q'}{A} \right) = 0.18 \times h_{fg} \times \rho_v \times \frac{\sigma \times g \times (\rho_1 - \rho_v)}{\rho_v^2}}$$

$$= 0.18 \times 2256.9 \times 10^3 \times 0.597 \times \left[\frac{0.0588 \times 9.81 \times (961 - 0.597)}{(0.597)^2} \right]^{0.25}$$

$$= 1.52 \times 10^6 \ w/m^2$$

> Critical heat flux $(Q'/A) = 1.52 \times 10^6 \ w/m^2$.

Step 7

Result

1. Heat flux, $Q/A = 142.83 \times 10^3 \ w/m^2$

2. Critical heat flux, $Q'/A = 1.52 \times 10^6 \ w/m^2$.

3.9. Problems Based On Condensation for Vertical Surfaces (Vertical Flatplate and Vertical Tube)

A vertical flat plate in the form of fin in 500mm in height and is exposed to stream at atmospheric pressure. If surface of the plate is maintained at 60°C. Calculate the following.

1. The film thickness at the trailing edge.

2. Overall heat transfer coefficient

3. Heat transfer rate

4. The condenses the mass flow rate

Step 1

Length (or) Height $L = 500mm = 0.5m$

Surface temperature, $Tw = 60°C$

Step 2

To find

1) δ_x 2) h 3) Q 4) $\dot{m}$

Step 3

We know that, Saturation temperature of water is 100°C,

> i.e., $T_{sat} = 100°C$

Properties of steam at 100°C, {from R.S. Khurmi Steam table, p.no.4}

$h_{fg} = 2256.9 \ KJ/kg$

$$h_{fg} = 2256.9 \times 10^3 \ \text{J/kg}.$$

We know that,

$$\text{film temperature, } T_f = \frac{T_w + T_{sat}}{2}$$

$$= \frac{60 + 100}{2}$$

$$T_f = 80^\circ C$$

Properties of Saturated Water at 80°C

{From HMTDB P.No 22}

$\rho = 974 \ \text{Kg/m}^3$

$v = 0.364 \times 10^{-6} \ \text{m}^2/\text{s}$

$k = 0.6687 \ \text{w/m.k}$

$\mu = \rho \times v = 974 \times 0.364 \times 10^{-6}$

$\mu = 354.33 \times 10^{-6} \ \dfrac{N-S}{m^2}$

Step 4

1) Film thickness (δ_x)

We know that,

For vertical plate

$$\delta_x = \left[\frac{4\mu k \ \times (T_{sat} - T_w)}{g \times h_{fg} \times \rho^2} \right]^{0.25}$$

{HMTDB P.No 149}

where,

x = L = 0.5m

$$\delta_x = \left[\frac{4 \times 354.53 \times 10^{-6} \times 0.6687 \times 0.5 \times (100 - 60)}{9.81 \ \times 2256.9 \times 10^3 \times (974)^2} \right]^{0.25}$$

$$\delta_x = 1.73 \times 10^{-4} \ \text{m}$$

Step 5

Average heat transfer coefficient (h)

For vertical surface, laminar flow.

$$h = 0.943 \left[\frac{k^3 \times \rho^2 \times g \times h_{fg}}{\mu \times L \times (T_{sat} - T_w)} \right]^{0.25}$$

The factor 0.943 may be replaced by 1.13 for more accurate result as suggested by MC Adams,

$$h = 1.13 \times \left[\frac{(0.6687)^3 \times (974)^2 \times 9.81 \times 2256.9 \times 10^3}{354.53 \times 10^{-6} \times 0.5 \times (100 - 60)} \right]^{0.25}$$

$$h = 6164.3 \ W/m^2.k$$

Step 6

Heat transfer rate (Q)

$$Q = hA(T_{sat} - T_w)$$

$$= h \times L \times w \times (T_{sat} - T_w) = 6164.3 \times 0.5 \times 1 \times (100 - 60)$$

$$Q = 123286 W$$

Step 7

Condensate mass flow rate ($\dot{m}$)

$$Q = \dot{m} \times h_{fg}$$

$$\dot{m} = Q/h_{fg}$$

$$\dot{m} = \frac{123286}{2256.9 \times 10^3}$$

$$\dot{m} = 0.054 \ kg/s$$

Result

1. δ_x = 1.73 x 10⁻⁴m

   $\delta_x = 1.73 \times 10^{-4} m$

2. h = 6164.3 w/m².k

3. Q = 123286w

4. $\dot{m}$ = 0.054 kg/s.

2. Dry saturated steam at a pressure of 2.45 bar condenses on the surface of a vertical tube of a height in 1m. The tubes surface temperature is a kept at 117°C. Estimate the thickness of the condensate film.

Step 1

Given

Pressure, P	= 2.45 bar
Distance (or) height (x)	= 1m
Surface temperature (T_w)	= 117°C

Step 2

Thickness of the condensate film (δ_x)

Step 3

Properties of steam at 2.45 bar,

{from R.S khurmi steam table, P.No.10}

T_{sat} = 127°C

h_{fg} = 2183 KJ/Kg

hfg = 2183 x 10³ J/Kg

Step 4

We know that,

$$\boxed{\text{film temperature, } T_f = \frac{T_w + T_{sat}}{2}}$$

$$= \frac{117 + 127}{2}$$

$$T_f = 122°C$$

Properties of saturated water at 122°C = 120°C,　　　　　{HMTDB P.No 21}

ρ = 945 Kg/m³

v = 0.247 x 10⁻⁶ m²/s

k = 0.6850 w/m.k

$\mu = \rho \times v = 945 \times 0.247 \times 10^{-6} = 2.33 \times 10^{-4}$ N-S/m².

Step 5

For vertical surfaces, (Assume the film is laminar)

$$\boxed{\delta_x = \left[\frac{4\mu k\ x(T_{sat} - T_w)}{g \times h_{fg} \times \rho^2}\right]^{0.25}}$$

$$= \left[\frac{4 \times 2.33 \times 10^{-4} \times 0.6850 \times 1 \times (127-117)}{9.81 \ \times \ 2183 \ \times 10^{3} \times (945)^{2}} \right]^{0.25} = \left[\frac{6.384 \times 10^{-3}}{1.912 \times 10^{13}} \right]^{0.25}$$

$$\boxed{\delta_x = 1.35 \times 10^{-4} \text{m}}$$

Result

Thickness of the condensate film, $\delta_x = 1.35 \times 10^{-4}$m

3.10. Problem based on Condensate Horizontal Tubes

1. A tube of 2m length and 25mm outer diameter is to be condense saturated steam at 100°C. While the tube surface is maintained at 92°C. Estimate the average heat transfer coefficient and the rate of condensation of steam if the tube is kept horizontal. The steam condenses on the outside of the tube.

Step 1

Given

Tube length L	= 2m
Diameter D = 25mm	= 0.025m
Saturated steam temperature, T_{sat}	= 100°C
Tube surface temperature, T_w	= 92°C

Step 2

To Find

1. Average heat transfer coefficient (h)

2. Rate of Condensation ($\dot{m}$)

Step 3

Properties of steam at 100°C, {from R.S. Khurmi steam table, P.No.4}

h_{fg} = 2256.9kJ/kg

$$\boxed{h_{fg} = 2256.9 \times 10^3 \text{ J/kg.}}$$

We know that,

$$\boxed{\text{film temperature, } T_f = \frac{T_w + T_{sat}}{2}}$$

$$= \frac{92 + 100}{2}$$

$$\boxed{T_f = 90^{\circ}C}$$

Properties of saturated water at 90°C,

$\rho = 965 \ Kg/m^3$

$v = 0.310 \times 10^{-6} m^2/s$

$k = 0.677 w/m.k$

$\mu = \rho \times v = 965 \times 0.310 \times 10^{-6}$

$$\boxed{\mu = 2.99 \times 10^{-4} N\text{-}S/m^2.}$$

Step 4

For horizontal tube, heat transfer coefficient,

$$\boxed{h = 0.728 \times \left[\frac{k^3 \ \rho^2 \ g \ h_{fg}}{\mu D \left(T_{sat} - T_w \right)} \right]^{0.25}}$$

$$h = 0.728 \left[\frac{(0.677)^3 \times (965)^2 \times 9.81 \times 2256.9 \times 10^3}{2.99 \times 10^{-4} \times 0.025 \times (100 - 92)} \right]^{0.25}$$

$h = 13166.08 \ w/m^2.k$

Average heat transfer coefficient = $13166.08 \ w/m^2.k$

Step 5

$$\boxed{\text{Heat transfer, } Q = hA \ (T_{sat} - T_w)}$$

$= h \times \pi \times D \times L \ (T_{sat} - T_w)$

$= 13166.08 \times \pi \times 0.025 \times 2 \times (100 - 92)$

$$\boxed{Q = 16544.98 W}$$

Step 6

We know that,

$Q = \dot{m} \times hfg$

$$\boxed{\dot{m} = Q/hfg}$$

$$\dot{m} = \frac{16544.98}{2256.9 \times 10^3}$$

$\overset{\bullet}{m}$ = 7.33 x 10⁻³kg/S

Rate of condensation, $\overset{\bullet}{m}$ = 7.33 x 10⁻³ kg/S.

Step 7

h = 13166.08 w/m².k

$\overset{\bullet}{m}$ = 7.33 x 10⁻³ kg/S.

2. A condenser is to be designed to condense 600kg/h of dry saturated steam at a pressure of 0.12 bar. A square array of 400 tubes, each of the 8cm diameter to be used. The tube surface is maintained at 30°C. Calculate the heat transfer coefficient and the length of the each tube.

Step 1

Given Data

$$\overset{\bullet}{m} = 600kg/h = \frac{600}{3600}\,kg/s \quad = 0.166\ kg/S$$

Pressure, P	= 0.12 bar
No. of tubes	= 400
Diameter, D = 8mm	= 8 x 10⁻³m
Surface temperature, Tw	= 30°C

Step 2

To Find

1. Heat Transfer Coefficient (h)
2. Length of the tubes (L)

Step 3

Properties of steam at 0.12 bar (from R.S Khumi steam table P.No.7)

T_{sat} = 49.45°C

h_{fg} = 2384.3kJ/kg

h_{fg} = 2384.3 x 10³ J/kg

We know that,

$$\text{film temperature, } T_f = \frac{T_w + T_{sat}}{2}$$

$$= \frac{30 + 49.45}{2}$$

$$T_f = 39.72°C = 40°C$$

Properties of saturated water at 40°C,

$\rho = 995 \ Kg/m^3$

$v = 0.657 \times 10^{-6} m^2/s$

$k = 0.628 w/m.k$

$\mu = \rho \times v = 995 \times 0.657 \times 10^{-6}$

$\mu = 653.7 \times 10^{-4} N\text{-}S/m^2.$

Step 4

with 400 tubes, a 20 x 20 tube square all could be formed

$N = \sqrt{400} = 20$

$$\boxed{N = 20}$$

For horizontal bank of tubes, heat transfer coefficient,

$$h = 0.728 \times \left[\frac{k^3 \ \rho^2 \ g \ h_{fg}}{\mu ND(T_{sat} - T_w)} \right]^{0.25} \qquad \{\text{HMT DB P. No. 149}\}$$

$$h = 0.728 \left[\frac{(0.628)^3 \times (995)^2 \times 9.81 \times (2384.3) \times 10^3}{653.7 \times 10^{-6} \times 20 \times 8 \times 10^{-3}(49.95 - 30)} \right]^{0.25}$$

$$\boxed{h = 53.04.75 \ w/m^2.k}$$

Step 5

Heat transfer, $Q = hA \ (T_{sat} - T_w)$

No. of tubes $= 400 = 400 \ \times h \times A \times (T_{sat} - T_w)$

$$= 400 \times 5304.75 \times \pi \times 8 \times 10^{-3} \times 2 \times (49.95 - 30)$$

Step 6

We know that, $\hspace{8cm}$ (1)

$$\boxed{Q = 1.05 \times 10^6 \times L}$$

$Q = \dot{m} \times hfg$

$\dot{m} = 0.166 \times 2384.3 \times 10^3$

$$\boxed{Q = 0.395 \times 10^6 w}$$

(2)

Step 7

Equating (1) and (2) to get L value,

$0.3957 \times 10^6 = 1.05 \times 10^6 L$

$L = 0.37m$

Step 8

Result

Heat Transfer Coefficient (h) = 5304. 75 $w/m^2.K$

Length of the tubes (L) = 0.37m

3.11. Application of Boiling and Condensation

1. Thermal & nuclear power plant
2. Refrigerating systems
3. Air conditioning systems
4. Heating of metal in fumaces
5. Process of heating & cooling

3.12. Heat Exchangers

Section I

Introduction

- Heat exchangers are devices that facilitate the exchange of heat between two fluids that are at different temperatures.
- Heat transfer in a heat exchanger usually involves convection in each fluid and conduction through the wall separating the two fluids.
- Thus in the analysis of heat exchangers, overall heat transfer co-efficient 'U' is used. Which accounts for all modes of heat transfer in heat exchangers.
- The rate of heat transfer between two fluids depends upon the magnitude of the temperature difference, which varies along the length of heat exchanger.
- Some common example of heat exchangers.
    - Condenser and evaporators.
    - Pregenerator

- Condenser and boiler in power plants
- Automobile radiators
- Inter coolers and preheaters
- Milk chiller of a pasteurising plant.

3.13. Classification and Application of Heat Exchangers

Heat exchangers are classified on basis of nature of heat exchange process, relative direction of fluid motion, design and constructional features and physical state of fluid.

(1) Nature of Heat Exchange Process

According to nature of heat exchange are classified into,

(i) Direct contact heat exchanger

(ii) Indirect contact heat exchanger

(i) In direct contact heat exchanger, both fluid mix with each other, thus heat and mass transfer takes place simultaneously.

This type of heat exchange is used only when mixing of both fluid is either harmless or desirable.

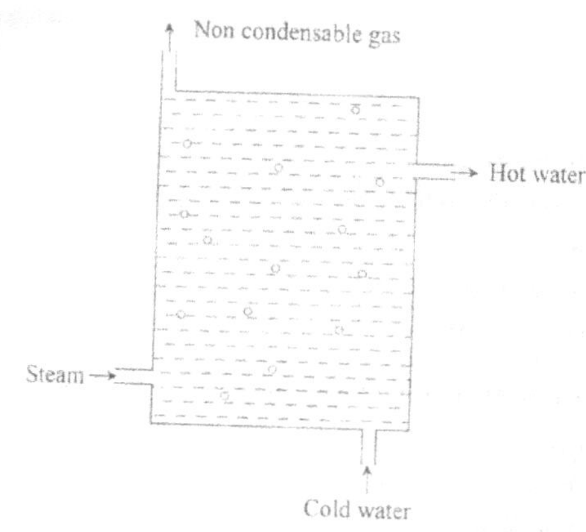

Fig. 3.4: Direct Contact Heat Exchanger

Application

- Jet condensers
- Cooling towers
- Direct contact feed heaters.

(ii) In indirect contact heat exchanger both fluid are separated by a thin wall and heat exchange takes place between two fluids through this wall.

Indirect contact heat exchangers are further classified into,

(a) Regenerator

(b) Reoperator or surface exchangers

(a) Regenerator

In a generator type heat exchanger, the hot and cold fluid flows alternately through space containing solid particles called matrix, this matrix act as sink and source for heat flow alternately.

Application

- Ic engine and gas turbines
- Open hearth and glass melting furances
- Air heaters of blast furnace.

The solid matrix alternately stores heat extracted from hot fluid and then delivers it to the cold fluid, thus regenerator operates periodically.

Heat exchange process can be made continuous if the matrix is made to rotate through fluid passages arranged side by side.

(b) Recuperators

In recuperator heat exchangers, both fluid are separated by a thin wall (in the form of pipe or tube). These heat exchangers are used when mixing of two fluid is undesirable. This is a most important type of heat exchanger.

Application

- Automobile radiator
- Milk chiller of pasteurising plant
- Evaporater of ice plant
- Oil coolers etc.

Advantages

Easy construction, more economical, more surface are for heat transfer

Disadvantages

Less heat transfer coefficient, less generating capacity.

(2) Relative Direction of Fluid Motion

According to relative direction of fluid motion heat exchangers are classified into

(i) Parallel flow heat exchangers

(ii) Counter flow heat exchangers

(iii) Cross flow heat exchangers.

(i) Parallel Flow Heat Exchangers

- In parallel flow heat exchangers both fluids (hot and cold) flows in the same direction.

- Both fluid enters at one end and leaves at the other end.

- Flow arrangement and temperature variation is shown in Fig. 3.5. the temperature difference between two fluids goes on decreases from inlet to outle as shown in Fig. 3.5.

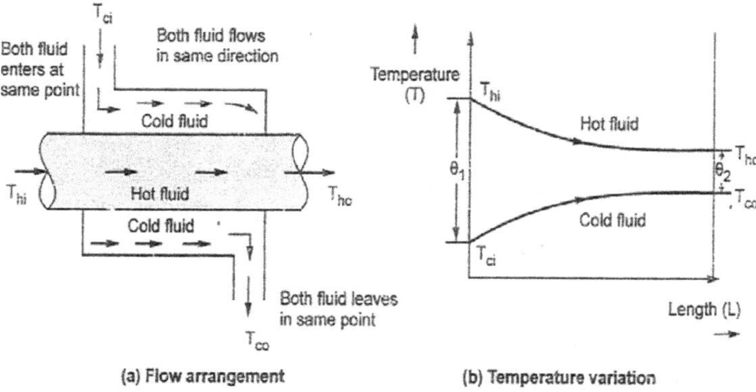

Fig. 3.5: Parallel Flow Heat Exchanger

- This type of arrangement requires the larger heat transfer, area thus it is rarely used in practice.

- Since both the fluids are separated by a tube, this type of heat exchangers are also called parallel flow recuperator or surface heat exchangers.

- Examples oil coolers, oil heaters, water heaters etc.

T_{hi} and $T_{ho} \rightarrow$ Inlet and out temperature of hot fluid.

T_{ci} and $T_{co} \rightarrow$ Inlet and out temperature of cold fluid

θ_1 and $\theta_2 \rightarrow$ Temperature difference at the inlet and outlet.

$\theta_1 = T_{hi} - T_{ci}$; $\theta_2 = T_{ho} - T_{co}$

(ii) Counter Flow Heat Exchangers

- In counter flow heat exchangers both fluid flows in the opposite direction.
- Both fluid enters and leaves the heat exchanger at the opposite end.
- Flow arrangement and temperature variation is shown in Fig 3.6 the temperature difference between two fluids remains constant as shown in Fig. 3.6 (b).
- The rate of heat transfer is maximum for same surface area as of parallel flow. Hence this type of arrangement is most widly used heating and cooling applications.

T_{hi} and $T_{ho} \rightarrow$ Inlet and out temperature of hot fluid.

T_{ci} and $T_{\infty} \rightarrow$ Inlet and out temperature of cold fluid

θ_1 and $\theta_2 \rightarrow$ Temperature difference at left and right side of heat exchanger.

$\theta_1 = T_{hi} - T_{co}$

$\theta_2 = T_{ho} - T_{ci}$

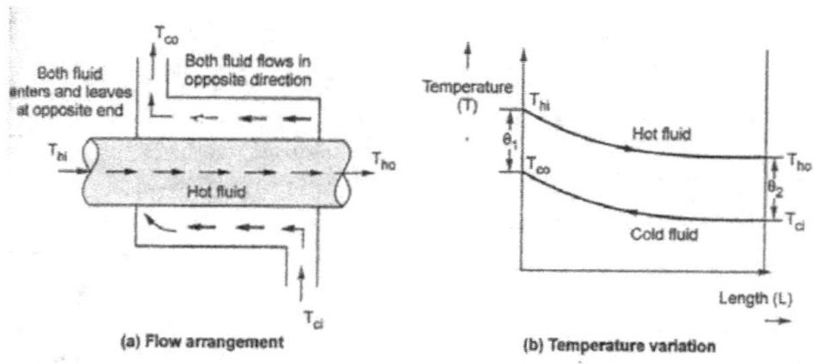

Fig. 3.6: Counter Flow Heat Exchangers

(iii) Cross-flow Heat Exchangers

- In Cross flow heat exchangers both fluids flows perpendicular to each other. The cross flow is further classified into unmixed and mixed flow, depending on flow configuration as shown in Fig. 3.7.

  (a) Unmixed cross flow heat exchanger, in this both fluids flow at right angle to each other and fluid flows in a well defined path (unmixed) as shown in Fig. 3.7 (a) .Ex : Automobile radiator.

  (b) Mixed cross flow heat exchanger, in this both fluids flow at right angle to each other. Referring to Fig. 3.7(b) only cold fluid is mixed and hot fluid is unmixed, thus temperature of cold fluid is uniform through the heat exchanger, but temperature of hot fluid varies along the flow direction.

  Ex : The cooling unit of refrigeration system.

There is a one more arrangement, in which both (hot and cold) fluid mix, such arrangement is called two fluid mixed cross flow heat exchanger. The temperature of both the fluid is uniform across the section and varies only in the direction of flow.

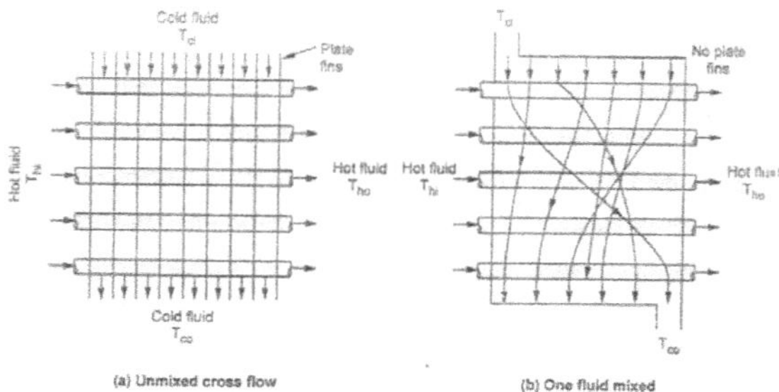

Fig. 3.7: Flow Arrangement for Cross Flow Arrangement

(3) Design and Constructional Features

According to design and constructional features, heat exchangers are classified into

(i) Compact heat exchangers

(ii) Concentric tube heat exchangers

(iii) Shell and tube heat exchangers

(iv) Multi – pass shell and tube passes.

(i) Compact Heat Exchangers

- Compact heat exchangers are special purpose heat exchangers and have a very large surface area per unit volume of the heat exchangers.

- Compact heat exchangers are generally employed when convective heat transfer coefficient associated with one of the fluid is much smaller than other fluid.

- Example : Plate fin, flattened fin tube exchangers. Etc.

(ii) Concentric Tube Heat Exchangers

In concentric tube heat exchanger, two concentric tubes are used, each carrying one of the fluid. The direction of flow may be parallel or counter.

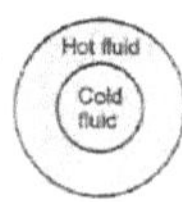

Fig. 3.8: Two Concentric Tube Heat Exchangers

(iii) Shell and Tube Heat Exchangers

- Shell and tube heat exchanger contain large number of tube packed in a shell, with their axes parallel to that of shell.
- Heat transfer takes place between two fluid, when one fluid flows though the shell and other through the tubes.
- Baffles are used in the shell to force the shell side fluid across the shell to enhance heat transfer and to maintain uniform spacing between the tubes.
- Shell and tube heat exchanger are not suitable to use in automotive and aircraft application because of large size and weight.

(iv) Multipass, Shell and Tube Heat Exchanger

- Shell and tube heat exchanger are further classified according to the number of shell and tube passes involved.
- Fig. 3.9 shows one shell and one tube pass shell and tube heat exchanger.
- If all the tubes make one U – turn in the shell, theses heat exchangers are called one shell pass and two tube pass heat exchangers.

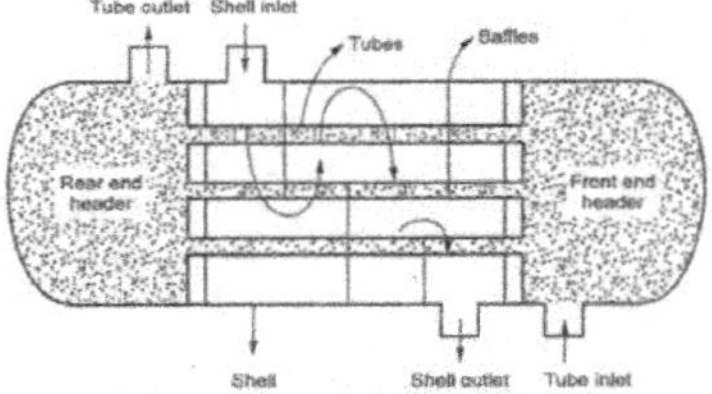

Fig. 3.9: Shell and Tube Heat Exchanger (One Shell and One Tube Pass)

- A heat exchanger that involves two passes in the shell and four passes in the tube is called a two – shell pass and four – tube pass heat exchanger.

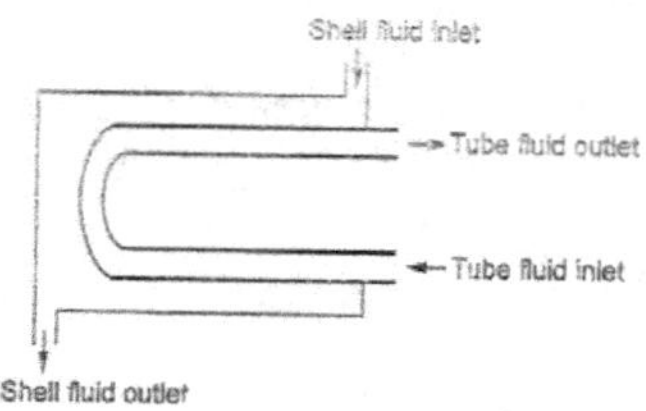

Fig. 3.10: One Shell Pass and Two Tube Pass Heat Exchanger

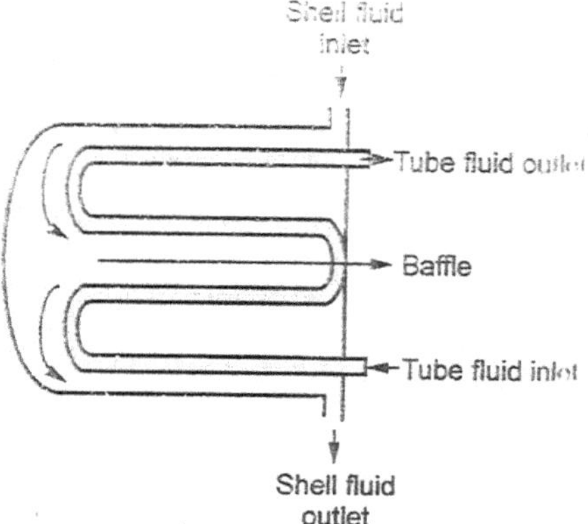

Fig. 3.11: Two Shell Pass and Four Tube Pass Heat Exchanger

(4) Physical State of Fluid

According to physical state of fluid, shell and tube heat exchanger are classified into

(i) Condensers

(ii) Evaporators.

Condensers

- In Condensers, the condensing fluid remains at the same temperature through the heat exchanger while the temperature of the cooler fluid gradually increases from inlet to outlet.

- The condensing fluid (hot fluid) losses latent heat which is received by cold fluid.

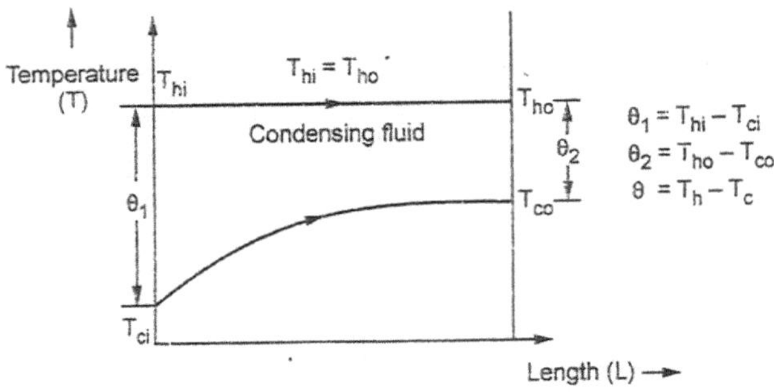

Fig. 3.12: Temperature Distribution in a Condenser

- In evaporators, the boiling fluid remains at the same temperature through the heat exchanger, while the temperature of the other fluid (hot) gradually decreases from inlet to outlet.

- The hot fluid losses heat of evaporation and it is received by boiling (cold) fluid.

- Fig. 3.13 shows the temperature variation for evaporator.

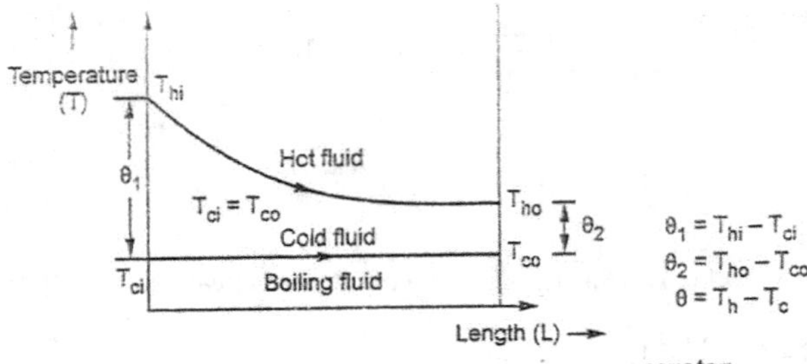

Fig. 3.13: Temperature Distribution in a Evaporator

3.14. Overall Heat Transfer Coefficient

- Let us consider a plane wall of thickness 'L' and thermal conductivity 'K' as Shown in Fig. 7.10.

- A heat exchanger typically involves two fluids separated by a solid wall. Heat is transferred by convection to the wall, then by conduction through the wall and then by convection from the wall as shown in Fig. 3.14.

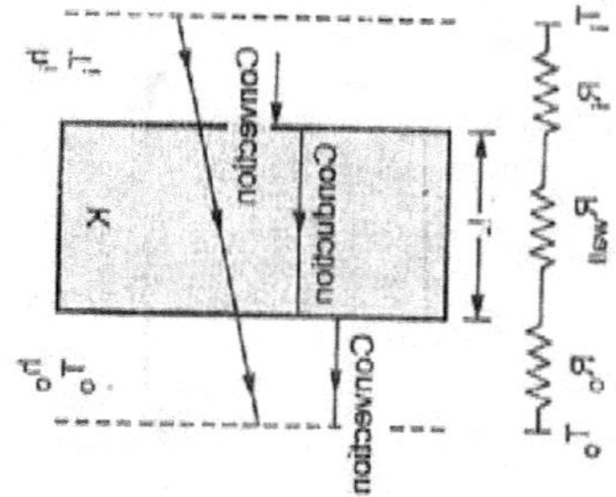

Fig. 3.14

- We have,

$$Q = \frac{\Delta T}{R} = \frac{T_i - T_o}{R_j + R_{wall} + R_o}$$

$$= UA_s \, (T_i - T_o)$$

$$= U_i A_i (T_i - T_o) = U_o A_o \, (T_i - T_o) \tag{3.1}$$

Where $U \rightarrow$ over all heat transfer coefficient, which includes conduction, convection and radiation effect. The unit of 'U' is W/m^2K or W/m^2C as that of convective heat transfer coefficient.

- From equation 7.1, we get

$$\frac{1}{UA_s} = \frac{1}{U_i A_i} = \frac{1}{U_o A_o} = R = \frac{1}{h_i A_i} + R_{wall} + \frac{1}{h_o A_o}$$

$$\therefore UA_s = \frac{1}{\sum R} = \frac{1}{\dfrac{1}{h_i A_i} + R_{wall} + \dfrac{1}{h_o A_o}} = U_i A_i = U_o A_o$$

$$U_i = \frac{1}{\dfrac{1}{h_i} + R_{wall} \times A_i + \dfrac{A_i}{h_o A_o}} \tag{3.2}$$

$$\text{Similarly } U_o = \frac{1}{\dfrac{A_o}{h_i A_i} + R_{wall} A_o + \dfrac{1}{h_o}} \tag{3.3}$$

3.15. Logarithmic Mean Temperature Difference (LMTD)

We know that the temperature difference between the hot and cold in the heat exchanger varies from point to point. In addition various area of heat transfer are involved. Therefore, based on the concept of separate mean temperature difference, also called logarithmic mean temperature difference, the total heat transfer rate in the heat exchanger is passed as.

$Q = UA \, (\Delta T)m$

Where

U – Overall heat transfer co-efficient, W/m^2K

A – Area, m^2

$(\Delta T)m$ – Logarithmic mean temperature difference.

Assumptions

In order to dervive expression for LMTD for various types of heat exchangers the following assumptions are made.

1. Flow is steady

2. The overall heat transfer co-efficient is constant

3. The specific heats of both fluids are constant

4. The mass flow rate of both fluids are constant

5. Axial conduction along the tube is negligible.

6. The change in kinetic and potential entergies of the fluids are negligible.

3.16. (a) Logarithmic Mean Temperature Difference for Parallel Flow

A single pass parallel flow heat exchangers is shown in Fig. 3.15.

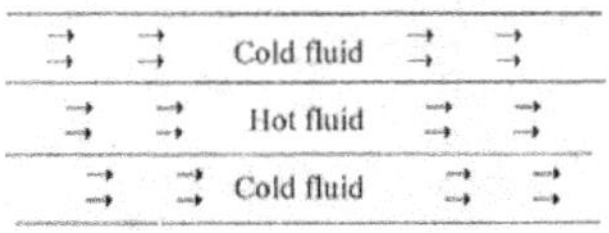

Fig. 3.15: Flow Arrangement

Let

m_h - mass flow rate of hot fluid.

m_c - mass flow rate of cold fluid.

C_{ph} - Specific heat of hot fluid.

C_{pc} - Specific heat of cold fluid.

T_1 – Entry temperature of hot fluid.

T_2 – Exit temperature of hot fluid.

t_1 - Entry temperature of Cold fluid.

t_2 - Exit temperature of Cold fluid.

U – overall heat transfer co-efficient.

Let us consider an elemental area dA of the heat exchangers the heat flow rate is given by

$$dQ = Uda\,(T-t) \tag{3.4}$$

We know that,

$$dQ = -m_h C_{ph} dT = m_c C_{pc} dt \tag{3.5}$$

$$\Rightarrow \quad dQ = -m_h C_{ph} dT$$

$$\Rightarrow \quad dT = \frac{-dQ}{m_h C_{ph}}$$

$$dT = \frac{-dQ}{C_h} \qquad [\because\ C_h = m_h \times C_{ph}] \tag{3.6}$$

From (3.15),

$$dQ = m_c C_{pc} dt$$

$$\Rightarrow dt = \frac{dQ}{Cc} \qquad [\because C_c = m_c \times C_{pc}] \qquad (3.7)$$

$$dT - dt = \frac{-dQ}{C_h} - \frac{dQ}{C_c}$$

$$= -dQ \left[\frac{1}{C_h} + \frac{1}{C_c} \right]$$

$$d\theta = -dQ \left[\frac{1}{C_h} + \frac{1}{C_c} \right] \qquad [\because d\theta = dT - dt] \qquad (3.8)$$

Substituting dQ value from Eqn. (3.4) in Eqn (3.8)

$$\Rightarrow \quad d\theta = -UdA\,(T\text{-}t) \left[\frac{1}{C_h} + \frac{1}{C_c} \right]$$

$$= -UdA\theta \left[\frac{1}{C_h} + \frac{1}{C_c} \right] \qquad [\because \theta = T - t]$$

$$\Rightarrow \quad \frac{d\theta}{\theta} = -UdA \left[\frac{1}{C_h} + \frac{1}{C_c} \right]$$

grating

$$\int_1^2 \frac{d\theta}{\theta} = -\left[\frac{1}{C_h} + \frac{1}{C_c} \right] U \int dA$$

$$\Rightarrow \quad [\ln \theta]_1^2 = -U \left[\frac{1}{C_h} + \frac{1}{C_c} \right] A$$

$$\Rightarrow \quad \ln\theta_2 - \ln\theta_1 = -UA \left[\frac{1}{C_h} + \frac{1}{C_c} \right]$$

$$\Rightarrow \quad \ln \left(\frac{\theta_2}{\theta_1} \right) = -UA \left[\frac{1}{C_h} + \frac{1}{C_c} \right] \qquad (3.9)$$

We know that,

$$Q = m_h C_{ph}\,(T_1 - T_2) = m_c C_{pc}\,(t_2 - t_1)$$

$$\Rightarrow \quad Q = C_h(T_1\text{-}T_2) = C_c(t_2\text{-}t_1) \qquad (3.10)$$

$$[\because \ C = m \times C_p]$$

$$\Rightarrow \quad Q = C_h(T_1 - T_2)$$

$$\Rightarrow \quad \frac{1}{C_h} = \frac{T_1 - T_2}{Q} \tag{3.11}$$

From eqn (3.10),

$$Q = C_c (t_2 - t_1)$$

$$\Rightarrow \quad \frac{1}{C_c} = \frac{t_1 - t_2}{Q} \tag{3.12}$$

Substitute $\dfrac{1}{C_h}$ and $\dfrac{1}{C_c}$ values in Eqn (3.9)

$$(3.9) \Rightarrow \ln\left(\frac{\theta_2}{\theta_1}\right) = -UA\left[\frac{T_1 - T_2}{Q} + \frac{t_2 - t_1}{Q}\right]$$

$$\Rightarrow \quad \ln\left(\frac{\theta_2}{\theta_1}\right) = -\frac{UA}{Q}\left[T_1 - T_2 + t_2 - t_1\right]$$

$$\Rightarrow \quad Q = \frac{-UA\left[T_1 - T_2 + t_2 - t_1\right]}{\ln\left(\dfrac{\theta_2}{\theta_1}\right)}$$

$$= \frac{UA\left[T_2 - T_1 + t_1 - t_2\right]}{\ln\left(\dfrac{\theta_2}{\theta_1}\right)}$$

$$= \frac{UA\left[T_2 - t_2 - (T_1 - t_1)\right]}{\ln\left(\dfrac{\theta_2}{\theta_1}\right)}$$

$$Q = \frac{UA\left[(T_2 - t_2) - (T_1 - t_1)\right]}{\ln\left(\dfrac{T_2 - t_2}{T_1 - t_1}\right)} \qquad\qquad [\because \ \theta = T - t]$$

$$Q = \frac{UA\left[(T_1 - t_1) - (T_2 - t_2)\right]}{\ln\left(\dfrac{T_1 - t_1}{T_2 - t_2}\right)} \tag{3.13}$$

(or)

$$Q = UA \ (\Delta T)m$$

Where $(\Delta T)m$ – Logarithmic mean temperature difference

$$(\Delta T)m = \frac{[(T_1 - t_1) - (T_2 - t_2)]}{\ln\left(\dfrac{T_1 - t_1}{T_2 - t_2}\right)}$$

(b) Logarithmic Mean Temperature Difference for Counter Flow

A single pass counter flow heat exchanger is shown in Fig. 3.16.

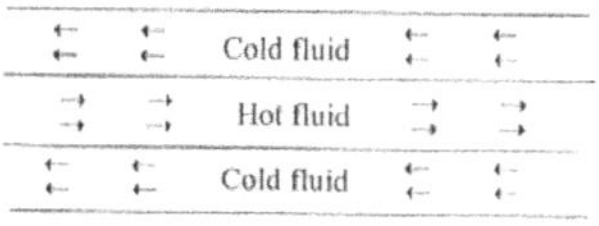

Fig. 3.16: Flow Arrangement

Let,

m_h	-	mass flow rate of hot fluid.
m_c	-	mass flow rate of cold fluid.
C_{ph}	-	Specific heat of hot fluid.
C_{pc}	-	Specific heat of cold fluid.
T_1	–	Entry temperature of hot fluid.
T_2	–	Exit temperature of hot fluid.
t_1	-	Entry temperature of Cold fluid.
t_2	-	Exit temperature of Cold fluid.
U	–	overall heat transfer co-efficient.

Let us consider an elemental area dA of the heat exchanger

The heat flow rate is given by

$$dQ = UdA\,(T-t) \tag{3.17}$$

We know that,

$$dQ = -m_h C_{ph}(dT) = -m_c C_{pc}\,(dt) \tag{3.18}$$

$$\Rightarrow \quad dQ = -m_h C_{ph} dT$$

$$\Rightarrow \quad dT = \frac{-dQ}{m_h C_{ph}}$$

$$dT = \frac{-dQ}{C_h} \tag{3.19}$$

From Equn (3.18),

$$[\because C_h = m_h \times C_{ph}] \qquad dQ = -m_c C_{pc} dt$$

$$\Rightarrow \quad dt = \frac{-dQ}{m_c C_{pc}}$$

$$dt = \frac{-dQ}{C_c} \tag{3.20}$$

$$[\because C_c = m_c \times C_{pc}]$$

$$dT - dt = \frac{-dQ}{C_h} + \frac{dQ}{C_c}$$

$$= -dQ\left[\frac{1}{C_h} - \frac{1}{C_c}\right]$$

$$d\theta = -dQ\left[\frac{1}{C_h} - \frac{1}{C_c}\right] \tag{3.21}$$

$$[\because d\theta = dT - dt]$$

Substituting dQ value from Equn (3.17) in Equn (3.21)

$$d\theta = -\,UdA\,(T - t)\left[\frac{1}{C_h} - \frac{1}{C_c}\right]$$

$$= -UdA\,\theta\left[\frac{1}{C_h} - \frac{1}{C_c}\right] \qquad [\because \theta = T - t]$$

$$\Rightarrow \quad \frac{d\theta}{\theta} = -UdA\left[\frac{1}{C_h} - \frac{1}{C_c}\right]$$

Grating

$$\int_1^2 \frac{d\theta}{\theta} = -\left[\frac{1}{C_h} - \frac{1}{C_c}\right]U\int dA$$

$$\Rightarrow \quad [\ln\theta]_1^2 = -UA\left[\frac{1}{C_h} - \frac{1}{C_c}\right]$$

$$\Rightarrow \quad \ln\theta_2 - \ln\theta_1 = -UA\left[\frac{1}{C_h} - \frac{1}{C_c}\right]$$

$$\Rightarrow \quad \ln\left(\frac{\theta_2}{\theta_1}\right) = -UA\left[\frac{1}{C_h} - \frac{1}{C_c}\right] \qquad (3.22)$$

We know that,

$$Q = m_h C_{ph}\,(T_1 - T_2) = m_c C_{pc}\,(t_2 - t_1)$$

$$\Rightarrow \quad Q = C_h(T_1 - T_2) = C_c\,(t_2 - t_1) \qquad (3.23)$$

$$\Rightarrow \quad Q = C_h\,(T_1 - T_2)$$

$$\Rightarrow \quad \frac{1}{C_h} = \frac{T_1 - T_2}{Q} \qquad (3.24)$$

From Equn (3.23)

$$Q = C_c\,(t_2 - t_1)$$

$$\Rightarrow \quad \frac{1}{C_c} = \frac{t_2 - t_1}{Q} \qquad (3.25)$$

Substitute $\dfrac{1}{C_h}$ and $\dfrac{1}{C_c}$ values in Equn (3.22)

$$\ln\left(\frac{\theta_2}{\theta_1}\right) = -UA\left[\frac{(T_1 - T_2)}{Q} - \frac{(t_2 - t_1)}{Q}\right]$$

$$\ln\left(\frac{\theta_2}{\theta_1}\right) = -\frac{UA}{Q}[(T_1 - T_2) - (t_2 - t_1)]$$

$$\ln\left(\frac{\theta_2}{\theta_1}\right) = -\frac{UA}{Q}[(T_1 - t_2) - (T_2 - t_1)]$$

$$\ln\left(\frac{\theta_2}{\theta_1}\right) = \frac{UA}{Q}[(T_1 - t_1) - (T_1 - t_2)]$$

$$\Rightarrow \quad Q = \frac{UA\,[(T_1 - t_2) - (T_2 - t_1)]}{\ln\left(\dfrac{\theta_2}{\theta_1}\right)}$$

$$\Rightarrow \quad Q = \frac{UA\,[(T_1 - t_2) - (T_2 - t_1)]}{\ln\left(\dfrac{T_2 - t_1}{T_1 - t_2}\right)} \qquad \begin{aligned}[\because \theta_2 &= T_2 - t_1\\ \theta_1 &= T_1 - t_2]\end{aligned}$$

$$\Rightarrow \quad Q = \frac{UA[(T_1 - t_2) - (T_2 - t_1)]}{\ln\left[\dfrac{T_1 - t_2}{T_2 - t_1}\right]}$$

$$\Rightarrow \quad Q = UA\,(\Delta T)_m \tag{3.26}$$

Where

$(\Delta T)_m$ – Logarithmic mean temperature difference.

$$(\Delta T)_m = \frac{[(T_1 - t_2) - (T_2 - t_1)]}{\ln\left(\dfrac{T_1 - t_2}{T_2 - t_1}\right)}$$

3.17. Fouling Factors

The surfaces of a heat exchangers do not remain clean after it has been in use for sometime. The surfaces become fouled with soaling (or) deposits.

- The effect of these deposits affecting the value of overall heat transfer coefficient (U)
- This effect is taken care of by introducing an additional thermal resistance called the fouling resistance (R_f) which is given by as follows

$$U_{outer} = \frac{1}{\dfrac{1}{h_o} + R_{fo} + \dfrac{r_o}{k}\,ln\left(\dfrac{r_o}{r_i}\right) + \left(\dfrac{r_o}{r_i}\right)R_{fi} + \left(\dfrac{r_o}{r_i}\right)\left(\dfrac{1}{h_i}\right)}$$

$$U_{inner} = \frac{1}{\dfrac{1}{h_i} + R_{fi} + \dfrac{r_i}{k}\,ln\left(\dfrac{r_o}{r_i}\right) + \left(\dfrac{r_i}{r_o}\right)R_{fo} + \left(\dfrac{r_i}{r_o}\right)\left(\dfrac{1}{h_0}\right)}$$

3.18. Effectiveness by Using Number of Transfer Units (NTU)

Logarithmic Mean Temperature Difference : (LMTD)

When inlet and outlet conditions are specified. But when the problem is to determine the inlet and exit temperature of heat exchanger, effectiveness method is used.

Effectiveness

The ratio of actual heat transfer to the maximum possible heat transfer.

$$\text{Effectiveness } E = \frac{\text{Actual heat transfer}}{\text{Maximum Possible heat transfer}} = \frac{Q}{Q_{max}}$$

NTU

$$\text{Number of transfer of units (NTU)} = \frac{UA}{C_{min}}$$

3.19. Fouling Process

1. Precipitation
2. Sedimendation
3. Chemical reaction fouling
4. Corrosion fouling
5. Biological fouling
6. Freeze fouling

3.20. Parameter Affecting Fouling

1. Velocity
2. Temperature
3. Water chemistry
4. Tube material

3.21. Prevention of Fouling

1. Design of heat exchanger
2. Treatment of process system
3. By using cleaning system

3.22. Properties to be Considered for Selection of Materials for Heat Exchanger

1. Physical properties
2. Mechanical properties
3. Climatic properties
4. Chemical Environment
5. Service life
6. Reliability

3.23. Common Failures in Heat Exchanger

1. Checking of tubes either expected
2. Excessive transfer rates in heat exchanger
3. Increase the pump pressure to maintain throughout
4. Excessive temperatures in heat exchangers.

3.24. Formula Used in the LMTD Method & NTU Method

S.NO	DESCRIPTION	PARELLEL FLOW	COUNTER FLOW	CROSS FLOW
1		$$Q = UA(\Delta T)_{lm}$$ Hot fluid =(T1 , T2) Cold fluid = (t1 , t2)	$$Q = UA(\Delta T)_{lm}$$	$$Q = FUA(\Delta T)_{lm}$$
2	Logarithmic mean temperature difference (LMTD) $(\Delta T)_m$	$$(\Delta T)_{lm} = \frac{(T_1 - t_1) - (T_2 - t_2)}{\ln\left[\frac{T_1-t_1}{T_2-t_2}\right]}$$	$$(\Delta T)_{lm} = \frac{(T_1 - t_2) - (T_2 - t_1)}{\ln\left[\frac{T_1-t_2}{T_2-t_1}\right]}$$	$$(\Delta T)_{lm} = \frac{(T_1 - t_2) - (T_2 - t_1)}{\ln\left[\frac{T_1-t_2}{T_2-t_1}\right]}$$
		T1 – Entry temperature of hot fluid($^{\circ}$C) T2 – Exit temperature of hot fluid ($^{\circ}$C) t1 – Entry temperature of cold fluid($^{\circ}$C) t2 – Exit temperature of cold fluid($^{\circ}$C)	T1 – Entry temperature of hot fluid($^{\circ}$C) T2 – Exit temperature of hot fluid ($^{\circ}$C) t1 – Entry temperature of cold fluid($^{\circ}$C) t2 – Exit temperature of cold fluid($^{\circ}$C)	T1 – Entry temperature of hot fluid($^{\circ}$C) T2 – Exit temperature of hot fluid ($^{\circ}$C) t1 – Entry temperature of cold fluid($^{\circ}$C) t2 – Exit temperature of cold fluid($^{\circ}$C)
3	Heat transfer (Q)	$$Q = \dot{m}_c \rho_{pc}(t_2 - t_1)$$ $$Q = \dot{m}_h c_{ph}(T_1 - T_2)$$ $$Q = UA(\Delta T)_{lm}$$ $$Q = \dot{m} * h_{fg}$$ U – over all heat transfer coefficient ,W/m²k (Direct given (or) Inner and outer diameter (r_i, r_o) and heat transfer coefficient (h_i, h_o) $$\frac{1}{U} = \frac{r_i}{r_o} + \frac{1}{h_i} + \frac{1}{h_o}$$ HMT DB Pg.No 157 A – area surface πDL (m²) hfg - Enthalpy of evaporation J/kg ṁc – mass of flow rate of cold fluid kg/s ṁh – mass flow rate of hot fluid kg/s Ch – specific heat of hot fluid J/kgk Cc – specific heat of cold fluid J/kgk	$$Q = \dot{m}_c C_c(t_2 - t_1)$$ $$Q = \dot{m}_h c_{ph}(T_1 - T_2)$$ $$Q = UA(\Delta T)_{lm}$$ $$Q = \dot{m} * h_{fg}$$	$$Q = \dot{m}_c \rho_c(t_2 - t_1)$$ $$Q = \dot{m}_h c_h(T_1 - T_2)$$ $$Q = UA(\Delta T)_{lm} (counter\ flow)$$ $$Q = \dot{m} * h_{fg}$$ F correction factor HMT DB Pg.No 162 From R and P
4	Mass flow rate (water (or) oil	$$\dot{m} = \rho_o * A * C$$ ρ – Density kg/m³ C – velocity m/s A – area $\pi/4(D^2)$ m²		
5	Heat lost by hot fluid = heat lost by cold fluid Qh – Qc	Qh = Qc $m_h C_{ph}(T_1\text{-}T_2) = m_c C_c(t_2\text{-}t_1)$		

NTU Method

[Number transfer units]

NTU method is used to determine the inlet (or) outlet temperature of heat exchanger

NTU method, two inlet temperature is given. In let temperature of hot liquid (T1) and cold liquid (t1)

S.NO	DESCRIPTION	FORMULA	HMTDB Pg.No
1	Capacity rate of hot liquid (c) in W/k	$C = \dot{m}_h * C_{ph}$ ṁh – mass flow rate of hot liquid (kg/s) Cph – Specific heat of hot liquid (J/kgk)	152
2	Capacity raqte of cold liquid (c) in W/k	$C = \dot{m}_h * C_{ph}$ ṁh – mass flow rate of cold liquid (kg/s) Cph – Specific heat of cold liquid (J/kgk)	152
3	To determine C_{min} and C_{max}	C_{min} C_{max}	-
4	Number of transfer units (NTU)	$$NTU = \dfrac{UA}{C_{min}}$$ U – over all heat transfer co-efficient W/m²k A – Area (m²)	152
5	To determine effectiveness of heat exchanger (ε)	From graph (parallel , counter , shall and tubes) HMT DB Pg.No163-168 $$\dfrac{C_{min}}{C_{max}} - curve$$ NTU – X axis	
6	Maximum possible heat transfer (Qmax)	$Q_{max} = C_{min}\left(T_1 - t_1\right)$	152
7	Actual heat transfer rate (Q)	$Q = \varepsilon * Q_{max}$	152
8	To determine T2,t2	$Q = \dot{m}_c C_{pc}\left(t_2 - t_1\right)$ $Q = \dot{m}_c C_{pc}\left(T_1 - T_2\right)$	

3.25. Problems Based on LMTD Method

1. In a double pipe counter flow heat exchapter, 10000kg/hr of an oil having a specific heat of 2095 J/Kg-k is from 80°C to 50°C by 8000 kg/hr of water entering at 25°C. determine the heat exchanger area of an overall heat transfer coefficient of 300 $\dfrac{W}{m^2 - k}$. Take C_p for water as 4180 J/kg.K.

Step 1

Hot fluid – oil (T_1, T_2)

Cold fluid – water (t_1, t_2)

The mass flow rate of (Hot fluid)

$\dot{m} = 10000 \ kg/hr$

$\dot{m}_h = \dfrac{10000}{3600} \ kg/g$

$\dot{m} = 2.777 \ kg/s.$

Specific heat of oil, $C_{ph} = 2095 \ J/kg\text{-}k.$

Entry temperature of oil, $T_1 = 80°C$

Exit temperature of oil, $T_2 = 50°C$

Mass flow rate of water, (cold fluid)

$\dot{m}_c = 8000 \ kg/h = \dfrac{8000}{3600} \ kg/s$

$\dot{m}_c = 2.22 \ kg/s$

Entry temperature of water, $t_1 = 25°C$

Overall heat transfer coefficient, $U = 300 \ w/m^2.k$

Specific heat of water, $C_{pc} = 4180 \ J/kg.K.$

Step 2

To Find

Heat exchanger area, A

Step 3

Solution

Heat loss by oil (hot fluid) = heat gained by water (cold fluid)

$Q_h = Q_c$

$\Rightarrow m_h \, C_{p_h} \, (T_1\text{-}T_2) = m_c \, C_{p_c} \, (t_2\text{-}t_1)$

$\Rightarrow m_h \, C_{p_h} \, (T_1\text{-}T_2) = mc \, C_{p_c} \, (T_2\text{-}T_1)$

$\Rightarrow 2.777 \times 2095 \times (80\text{-}50) = 2.22 \times 4180x \ (t_2\text{-}25)$

$174.53 \times 10^3 = 9.27x \ 10^3 \, t_2 - 231.99 \times 10^3$

$t_2 = 43.85°C$

Exit temperature of water, $t_2 = 43\text{-}85°C.$

Heat transfer $Q = \dot{m}_h \ \ C_{p_h} \, (T_1\text{-}T_2)$

Or

$$\dot{m}_c \; C_{Pc}(t_2 - t_1)$$

Q = 2.22 x 4180 x (43.85-25)

Q = 174.92 x 10³W

Step 4

We know that,

Heat transfer Q = UA $(\Delta T)_m$

{HMT DB P.No 151}

$(\Delta T)_m$ = logarithmic mean temperature difference (LMTD)

For counter flow,

$$(\Delta T)_m = \frac{(T_1 - t_2) - (T_2 - t_1)}{\ln\left(\dfrac{T_1 - t_2}{T_2 - t_1}\right)}$$

$$= \ln \frac{(80 - 43.85) - (50 - 25}{\ln\left(\dfrac{80 - 43.85}{50 - 25}\right)}$$

$(\Delta T)_{lm}$ = 30.23°C

Sub $(\Delta T)_m$ and u and Q value in area 1.

Q = UA $(\Delta T)_m$

174.92 x 10³ = 300 x 4x 30.23

A = 19.287m²

Heat exchange area, A = 19.287 m².

2. In a counter flow double pipe heat exchanger, water is heated from 25°C to 65°C by an oil with a specific heat of 1.45 kJ/kg.k and the mass flow rate is 0.9 kg/s. the oil is cooled from 230°C to 160°C. If the overall heat transfer coefficient is 420 $W/m^2\ ^oC$. Calculate the following.

   1. The rate of heat transfer
   2. The mass flow rate of water
   3. The surface area of the heat exchanger.

Step 1

Hot fluid – oil (T_1, T_2) & cold fluid – water (t_1, t_2)

Entry temperature of water, $t_1 = 25°C$

Exit temperature of water, $t_2 = 65°C$

Specific heat of oil, $C_{p_h} = 145$ kJ/kg

$= 1.45$ x 10^3 J/kg.

Mass flow rate of oil,

$\dot{m}_h = 0.9$ kg/g

Entry temperature of oil, $T_1 = 230°C$

Exit temperature of oil, $T_2 = 160°C$

Overall heat transfer coefficient,

$U = 420$ W/m²°C.

Step 2

To find:

1. The rate of heat transfer, Q

2. Mass flow rate of water, $\dot{m}_C$

3. Surface area of the heat exchanger, A.

Step 3

Heat transfer, $Q = \dot{m}_h \, C_{p_h} \, (T_1 - T_2) = 0.9$ x 1.45 x 10^3 (230-160)

$$Q = 91.35 \text{ x } 10^3 \text{ W}$$

We know that,

Heat transfer, $Q = UA \, (\Delta T)l_m$

{HMT DB PN.O 157}

$(\Delta T)_m$ = logarithmic mean temperature

Difference (LMTD).

For counter flow,

$$(\Delta T)_m = \frac{[(T_1 - t_2) - (T_2 - t_1)]}{ln\left(\dfrac{T_1 - t_2}{T_2 - t_1}\right)} = \frac{(230 - 65) - (160 - 25)}{ln\left(\dfrac{230 - 65}{160 - 25}\right)}$$

$(\Delta T)_m = 149.49°C.$

Step 4

Sub, $(\Delta T)_m$ and Q& U values in can 1.

$Q = UA\,(\Delta T)_m$

$91.35 \times 10^3 = 420 \times A \times 149.49$

$$\boxed{A = 1.455\text{m}^2}$$

We know that, for cold fluid:

$Q = \dot{m}_c C_{Pc}\,(t_2 - t_1)$

$91.35 \times 10^3 = \dot{m}_c \times 4186 \times (65 - 25)$

[$\therefore$ Specific heat of water C_{Pc} = 4186 J/kg.k]

$\dot{m}_c = 0.545$ kg/s.

Result

1. Heat transfer $Q = 91.35 \times 10^3$ w

2. Mass flow rate of water, $\dot{m}_c = 0.545$ kg/s

3. Surface area of the Heat Exchanger, $A = 1.455$ m^2.

3. A counter flow concentric tubes heat exchanger is used to cool engine oil (c=2130 J/kg.k) from 160°C to 60°C with water available at 25°C as the cooling medium. The flow rate of cooling water through the inner tube of 0.5m s 2kg/s while the flow rate of oil through the outer annulus. Outer diameter = 0.7m is also 2 kg/s. if u is 250 w/m^2.k, how long must the heat exchanger to meat the cooling requirement?

Step 1

Given

Hot fluid oil –   cold fluid water

(T_1, T_2) (t_1, t_2)

Specific heat of the oil (Hot fluid) = C_{P_h} = 2130 J/Kg.k

Entry temperature of oil, $T_1 = 160$°C

Exit temperature of oil, $T_2 = 60$°C

Entry temperature of water, $t_1 = 25$°C

Inner diameter, $D_1 = 0.5$m.

Flowrate of water (cooling fluid)

Outer diameter $D_2 = 0.7$m

Flow rate of oil (Hot fluid), $\dot{m}_h = 2$kg/s

Overall heat transfer coefficient, $U = 250$ w/m².k.

Step 2

Length of the heat exchanger, L.

Step 3

We know that,

Heat lost by oil (Hot fluid) = heat seined by (cold fluid)

$\quad Q_h = Q_c$

$\Rightarrow \dot{m} \; C_{p_h} \; (T_1\text{-}T_2) = \dot{m}_C \; C_{p_c} \; (t_2\text{-}t_1)$

$= 2 \times 2130 \times (160\text{-}60) = 2 \times 4186 \times (t_2 - 25)$

$42.6 \times 10^4 = 8.372 \, (t_2\text{-}25)$

$t_2 = 75°88°C.$

Heat transfer, $Q = \dot{m}_h \; C_{p_h} \; (t_2\text{-}t_1)$ (or) $\dot{m} \; C_{p_h} \; (T_1\text{-}T_2)$

$Q = 2 \times 4186 \times (75.88\text{-}25)$

$Q = 425.96 \times 10^3$w.

Step 4

Heat transfer, $Q = UA \, (\Delta T)_m$

Where $(\Delta T)_m =$ logarithmic mean temperature difference (LMTD)

Counter flow,

$$(\Delta T)_m = \frac{\left[(T_1 - t_2) - (T_2 - t_1)\right]}{\ln\left(\dfrac{T_1 - t_2}{T_2 - t_1}\right)}$$

$$(\Delta T)_m = \frac{(160 - 75.88) - (60 - 25)}{\ln\left(\dfrac{160 - 75.88}{60 - 25}\right)}$$

$$= \frac{49.12}{0.8768}$$

$(\Delta T)_m = 56.02°C$

Step 5

Sub $(\Delta T)_m$, u, Q values sub in can 1.

$Q = UA\,(\Delta T)_m$

$425.96 \times 10^3 = 250 \times A \times 56.02°C$

$$\boxed{A = 30.415 \text{ m}^2.}$$

Step 6

We know that,

$$\boxed{A = \pi \times D_1 \times L}$$

Area $= \pi \times 0.5 \times L$

$L = 19.36$ m.

Length of heat exchange, (Single pass)

$$\boxed{L = 19.36\text{m.}}$$

Step 7

Result

Length of the heat exchanger, L=19.36m.

3.26. Problems based on the NTU Method

1. It is designed to use double pipe counter flow heat exchanger to cool 3 kg/s of oil ($c_p = 2.1$ J/kg.k) from 120°C. cooling water at 20°C. enters the heat exchanger at a rate of 10kg/s. the overall heat transfer area is 6m².

    Calculate the exit temperatures of oil and water.

Step 1

Cold fluid – water (t_1, t_2)

Hot fluid - oil (T_1, T_2)

Mass flow rate of oil, $\overset{\bullet}{m}_h$ $= 3$ kg/g

Specific heat of oil, $C_{p_h} = 2.1$ kJ/kg.k

$C_{p_h} = 2.1 \times 10^3$ J/kg.k

Inlet temperature of oil = T_1 = 120°C.

Inlet temperature of water, t_1 = 20°C

Mass flow rate of water $\dot{m}_c$ = 10 kg/s

Overall heat transfer

Coefficient, (U) = 600 w/m².k

Heat exchanger area, A = 6 m²

Step 2

1. Exit temperature of oil (T_2)
2. Exit temperature of water (t_2).

Step 3

Capacity rate of hot oil,

$$C = \dot{m}_h \times C_{Ph}$$

$$C = 3 \times 2.1 \times 10^3$$

$$C = 6300 \text{ w/k} \tag{1}$$

Capacity rate of water,

$$C = \dot{m}_c \times C_{pc}$$

$$C = 10 \times 4186 = 41860 \text{ w/k} \tag{2}$$

[∴ Specific heat of Water, C_{Pc} = 4186 w/kg.k]

Divided (1) by (2)

C_{min} = 6300 w/m.k

C_{max} = 41860 w/m.k

$$\frac{C_{min}}{C_{max}} = \frac{6300}{41860}$$

$$\boxed{\frac{C_{min}}{C_{max}} = 0.150} \tag{3}$$

Number of transfer of units (NTU) = $\dfrac{UA}{C_{min}}$ = $\dfrac{600 \times 6}{6300}$

NTU = 0.571 {To find effectiveness $\in$ in the HMT DB P.NO 163 for counter flow}

Step 3

For graph,

X-axis $\rightarrow$ NTU $\rightarrow 0.571$

Curve $\rightarrow \dfrac{C_{min}}{C_{max}} = 0.150$

Corresponding x axis value is 42 %

$(\varepsilon = 0.42)$

Step 4

Effectiveness

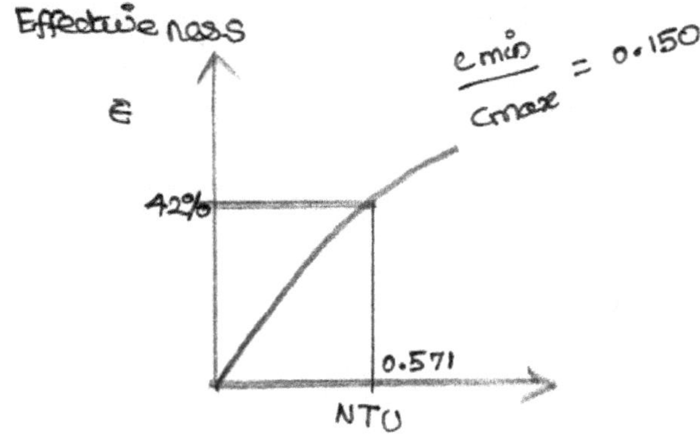

Maximum possible heat transfer,

Q_{max} = $C_{min}(T_1 - t_1)$ = 6300 (120-20)

Q_{max} = 63×10^4 W

Actual heat transfer rate,

Q = $\varepsilon \times Q_{max}$ = $0.42 \times 63 \times 10^4$

Q = 264600 W

We know that,

Q = $\dot{m}_c \, C_{Pc}(t_2 - t_1)$ 264600 = 10 x 4186 x (t$_2$-20)

t_2 = 26.32°C

Exit temperature of water, $t_2 = 26.32°C$.

Step 5

We know that,

Heat transfer Q = $\dot{m}_h \, C_{Ph} (T_1 - T_2)$

264600 = $3 \times 21 \times 10^3 (120 - T_2)$

T_2 = 78°C

Exit temperature of oil, $T_2 = 78°C$.

Step 6

1. T_2 = 78°C

2. t_2 = 26.32°C.

2. A Parallel flow heat exchanger has hot and cold water stream running through it, the flow rates are 10 and 25 kg/min respectively. Inlet temperatures are 75°C and 25°C on hot and cold sides. The exit temperature on the hot side should not exceed 50°C assume $h_i = h_o = 600 \text{ w/m}^2.\text{k}$. calculate the area of heat exchanger using ε–NTU method.

Step 1

Given

Mass flow rate of

Hot water $\dot{m}_h$ = 10 kg/min = $\dfrac{10}{60}$ kg/s

$\dot{m}_h$ = 0.166 kg/s

Maas flow rate of Cold water, $\dot{m}_C$ = 25 kg/min

= 25x60 kg/s = 0.416 kg/s.

Inlet temperature of hot Water, T_1 = 75°C

Inlet temperature of cold Water, t_1 = 25°C

Exit temperature of hot water, T_2 = 50°C

Overall heat transfer coefficient, h_o = h_i = 600 w/m².k.

Step 2

To Find

Heat exchange area, A

Step 3

Capacity rate of hot fluid,

$$C = \dot{m}_h \times C_{Ph}$$

$$C = 0.166 \times 4186$$

$$C = 694.87 \; w/k \qquad \qquad \rightarrow(1)$$

$[\because$ Specific heat of water, $C_p = 4186$ J/kg.k$]$

Capacity rate of cold fluid,

$$C = \dot{m}_C \times C_{Pc}$$
$$C = 0.416 \times 4186$$
$$C = 1741.37 \; w/k \qquad \qquad \rightarrow(2)$$

Step 4

Divided (1) by (2)

$$C_{min} = 694.87 \; w/k$$

$$C_{max} = 1741.37 \; w/k$$

$$\Rightarrow \frac{C_{min}}{C_{max}} = \frac{694.87}{1741.37}$$

$$\frac{C_{min}}{C_{max}} = 0.399 \qquad \qquad \rightarrow(3)$$

We know that,

Effectiveness, $\in = \dfrac{T_1 - t_2}{T_1 - t_1}$ $\qquad$ {HMTDB P.NO 151 }

$$= \frac{75 - 50}{75 - 25}$$

$$\in = 0.5$$

[To find NTU, Refer HMTDB P.No 163]

Step 5

From graph,

X axis $\rightarrow \in = 0.5$

Curve $\rightarrow \dfrac{C_{min}}{C_{max}} = 0.399$

Corresponding x axis value is 0.84 (ie, NTU = 0.84)

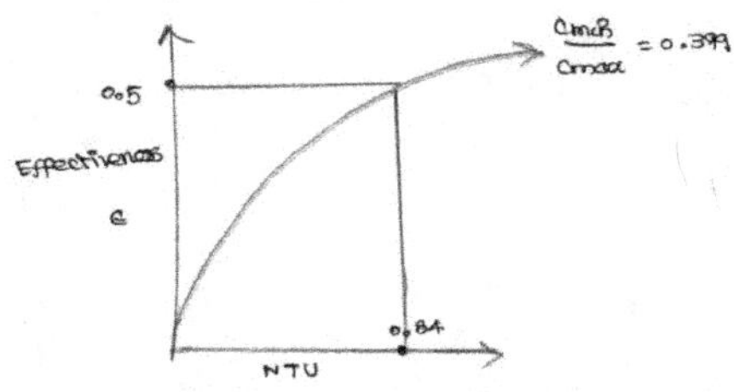

We know that,

NTU (Number of transfer units) $= \dfrac{UA}{C_{min}}$ {From HMTDB P.NO 152}

Step 6

Overall heat transfer coefficient, $\dfrac{1}{U} = \dfrac{1}{h_i} + \dfrac{1}{h_0}$

$$\dfrac{1}{U} = \dfrac{600 + 600}{600 + 600}$$

$$\dfrac{1}{U} = \dfrac{1}{300} \; \dfrac{1}{W/_{m^2.k}}$$

$$U = 300 \; W/_{m^2.k}$$

Substitute NTU, C_{min} and U values in equation (4),

$$(4) \Rightarrow NTU = \dfrac{UA}{C_{min}}$$

$$0.84 = \dfrac{300 \times A}{694.87}$$

$$A = 1.945 m^2.$$

Result

$$\boxed{\text{Heat exchange area, A} = 1.945 \ m^2.}$$

Unit 4

Radiation

4.1. Introduction

The heat is transferred from one body to another without any transmitting medium is known as radiation. It is an electromagnetic wave phenomenon.

4.2. Emission Properties

1. The wavelength (or) frequency of radiation
2. The temperature of surface
3. The nature of the surface

4.3. Emissive Power (Eb)

The Emissive power is defined as the total amount of radiation emitted by a body per unit time and unit area is denoted by (w/m^2)

4.4. Monochromatic Emissive Power (Ebλ)

The energy emitted by the surface at a given length per unit time per unit area in all directions is known as monochromatic emissive power.

4.5. Absorption, Reflection and Transmission

When incident radiation (also called irradiation impinges on a surface, three things happen: a part is reflected back, a part is transmitted through and the remainder is absorbed as shown in Fig. 4.1. if Q be the rate at which a surface receives heat and of this amount Q_p is reflected, Q_τ transmitted and Q_α absorbed, then by the principle of conservation of energy, the total sum must be equal to the incident radiation, i.e., reflection + transmission + absorption = incident radiation.

$$Q_p + Q_\tau + Q_\alpha = Q \text{ or } \frac{Q_p}{Q} + \frac{Q_\tau}{Q} + \frac{Q_\alpha}{Q} = 1$$

Or

$$\rho + \tau + \alpha = 1 \tag{4.1}$$

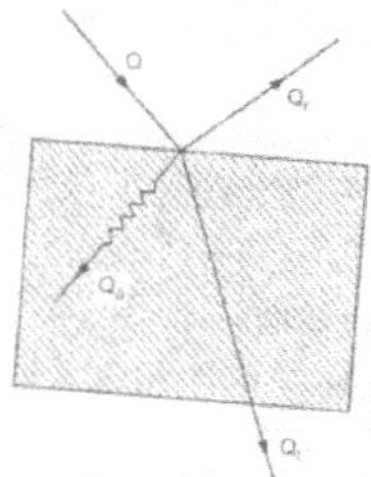

Fig. 4.1

Where $\rho = \dfrac{Q_\rho}{Q}$ is the fraction of incident radiation reflected and is called the reflectivity.

Similarly transmissivity, $\tau = \dfrac{Q_\tau}{Q}$ is defined as the fraction transmitted, and the absorptivity,

$\alpha = \dfrac{Q_\alpha}{Q}$ is the fraction absorbed.

Equation (4.1) holds for surfaces or layers of finite thickness. The following points should be noted about ρ, τ and α

1. They are always positive and their values lie between the limits 0 and 1, i.e., $0 \le \rho, \tau, \alpha \le 1$.

2. $\rho = 0$ (i.e., $\tau + \alpha = 1$) represents a non-reflecting surface; $\rho = 1$ (i.e., $\alpha + \tau = 0$) represents a perfect reflector, i.e., it reflects all the incident radiation and does not adsorb or transmit any part of it.

3. $\tau = 0$ (i.e., $\rho + \alpha = 1$) represents an opaque surface;

    $\tau = 0$ (i.e., $\rho = \alpha = 0$) represents a perfectly transparent surface.

4. $\alpha = 1$ (i.e., $\rho + \tau = 1$) represents a non-absorbing surface (also called a white surface);

    $\alpha = 1$ (i.e., $\rho = \tau = 0$) represents a perfectly absorbing surface (also called a black surface, if it is diffuse).

The definitions of ρ, τ and α given above are with respect to the total values of Q, integrated with respect to the area of the surface, the solid angle in hemispherical space above it and all the wavelengths of the spectrum. We can also define monochromatic and directional values of ρ, τ and α by taking the corresponding values of Q_ρ, Q_τ and Q_α

$$\rho\lambda = \frac{Q_{\rho\lambda}}{Q_\lambda}; \tau_\lambda = \frac{Q_{\tau\lambda}}{Q_\lambda}; \alpha_\lambda = \frac{Q_{\alpha\lambda}}{Q_\lambda} \tag{4.2}$$

Where Q_λ now represents the total heat flux per unit area received at the point at that wavelength and ρ_λ is the monochromatic reflectivity or the fraction of incident energy in the wavelength range λ to $\lambda + d_\lambda$, which is reflected.

Since

$$Q_\rho = \int_0^\infty \left(Q_{\rho\lambda} \right) d\lambda = \int_0^\infty \rho_\lambda Q_\lambda d_\lambda \qquad (4.3)$$

$$\rho = \frac{Q_\rho}{Q} = \frac{1}{Q} \int_0^\infty \rho_\lambda Q_\lambda d\lambda \qquad (4.4)$$

We can derive equations similar to Eqn. (4.4) for monochromatic absorptivity, α_λ, and monochromatic transmissivity, τ_λ, as well.

Solids generally do not transmit unless the material is of very thin section. Metals absorb radiation within a fraction of a micrometre, and insulators within a fraction of a millimetre. Glasses and liquids absorb most of the radiation within a millimetre. Solids and liquids are therefore generally considered as opaque.

Gases such as hydrogen, oxygen and nitrogen (and their mixtures such as air) have a transmissivity of practically unity. The radiation transfer through air is estimated using the relationships for radiation through a vacuum.

Two types of reflection phenomena-specular and diffuse are observed when radiation strikes a solid surface. Specular reflection occurs from a surface such as a mirror, which is very smooth. An image of the source of radiation is projected; the angle of reflection from the surface occurs practically indiscriminately in all diffractions. These two types reflection are shown in fig.4.2 no actual body is perfectly specular or diffuse but it is often to approximate to one these ideal surfaces.

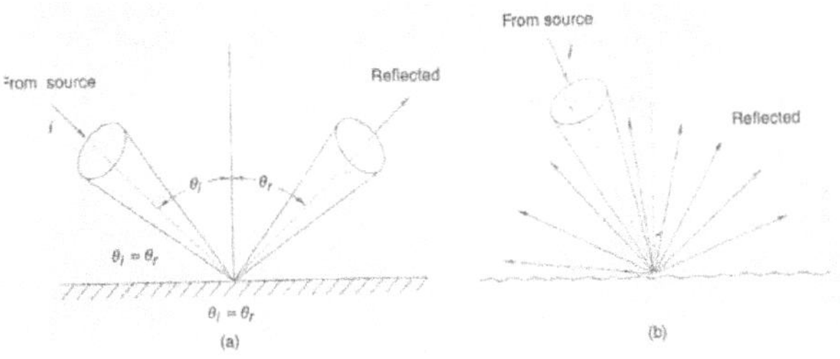

Fig. 4.2: Reflection (A) Specular, (B) Diffuse

4.6.   Concept of a Black Body

A black body is an ideal body that absorbs all incident energy and reflects or transmits none. This is true of radiation for all wavelengths and for all angles of incidence for a black body, therefore, $\rho=0$, $\tau = 0$ and $\alpha = 1$. We call it black becaouse the materials which obey this law appear black to this eye. It should be noted that surfaces which are nearly black for radiation purposes, are not necessarily black to visible to visible light becaoue the visible wavelength range is only a small part of the thermal radiation range. Snow and ice or white paper are quite bright to the eye but are nearly radiation black with an absorptivity of 0.97-0.98. No actual body is perfectly black; the concept of a black body is an idealisation with which the radiation characteristics of real bodies can be conveniently compared. A black body plays a role in thermal radiation similar to the idealized Carnot cycle in thermodynamics with which real cycles are compared.

A black body is regarded as a perfect absorber of incident radiation. A black condition can be approached in practice by forming a cavity in a material as shown in fig. 4.3. Radiation passing through the hole into the cavity is repeatedly absorbed and reflected at the cavity walls until it is all absorbed. The cavity itself may be made of a shiny material but will appear black when one looks in through the opening.

A black body is a perfect emitter. This is a fact which can be proved as follows. Consider a black body at a uniform temperature, placed inside an arbitrarily shaped, perfectly insulated enclosure composed of another black body whose temperature is also uniform but different from that of the former (Fig.4.4). The black body and the enclosure will reach a common equilibrium temperature after a period of time (due to heat transfer) when the black body will radiate as much energy as it absorbs; if this were not true, its temperature would change a direct violation of the second law of thermodynoamics. Becoause the black body is by definition absorbing the maximum possible radiation from the enclosure at each wavelength and from each direction, it must also be emitting the maximum total amount of radiation. An interesting point to note here is that a black body continues to emit radiation even when it is in thermal equilibrium with its surroundings.

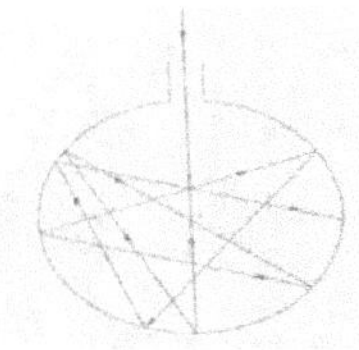

Fig. 4.3: Cavity Acting Like a Black Surface

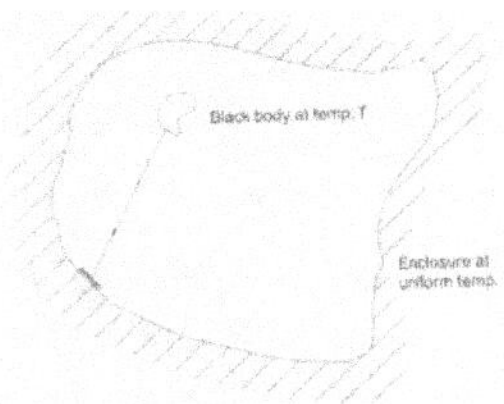

Fig. 4.4: Black Body in an Isothermal Enclosure

The total radiation emitted by a black is a function of temperature only. If the temperature of the enclosure in fig.4.4 is now changed to a different uniform value, the black body will adjust its temperature until is in thermal equilibrium with the enclosure. Then it will absorb and emit the maximum possible energy characteristic of its new temperature. Obviously, the rate of absorption and emission increases with the rise in temperature.

4.7.   Intensity of Radiation

A black body emitting radiation in all directions can be visualised by considering an elementary area, dA, of fig.4.5; the radiation will be wholly intercepted by the hemisphere of radius, R, in whose base dA lies. If dA1 is a corresponding elemental area on the normal from dA to the hemisphere, the solid angle it subtends at dA, dW, is dA_1/R^2. The solid angle (dW) is defined as the ratio of the area of the surface of a sphere enclosed by the conical surface forming the angle to the square of the radius of the sphere, i.e., $dW = dA_1/R^2$. The unit used is the steradian Sr. The solid angle of a hemisphere surrounding dA is $2\pi R^2/R^2 = 2\pi$ Sr. consider now the radiation emitted by dA which reaches a point P on an elementary area. dA_ρ, on the hemispherical surface displaced by an angle ϕ from the normal to the surface. Obviously, although radiation propagates from dA in all directions over the entire hemisphere, only a part which falls within the solid angle subtended by dA_ρ at dA will reach $dA\rho$.

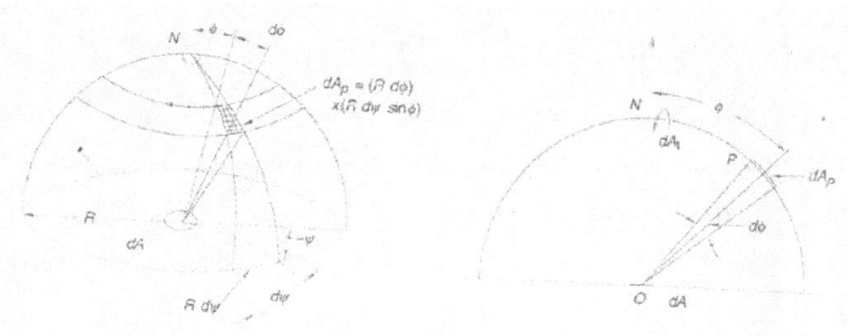

Fig. 4.5: Intensity of Radiation

The radiation emitted in any direction is defined in terms of the radiation intensity. The radiation intensity of a surface is defined as the rate of heat flux emitted by it per unit projected area normal to the direction of radiation per unit projected area normal to the direction of radiation per unit solid angle. If E_b at point O represents the sum of the total energy per unit area radiated in all possible directions of the hemispherical space, then it is called the total hemispherical or emissive power of the surface at the point O. if $dE_{b\phi}$ is the emissive power due to the energy radiated in solid angle, dW, in a particular direction ϕ, then the radiation intensity of the surface at O in that direction is

$$I_{b\phi} = \frac{dQ_b}{dA\cos\phi\,dW} = \frac{1}{\cos\phi}\left(\frac{dE_{b\phi}}{dW}\right) \qquad (4.5)$$

The reason for $\cos\phi$ coming into the above equation is that E_b is defined per unit area whereas $I_{b\phi}$ is defined per unit projected area. Equation (4.5) can be written in the integral form as

$$E_b = \int I_{b\phi}\cos\phi\,dW \qquad (4.6)$$

There are tow types of intensities - (1) The spectral radiation intensity, $Ib\lambda$, refers to radiation in an interval $d\lambda$ around a single wavelength propagating in a given direction at a position r. it represents the amount of energy streaming through a unit area projected normal to a given direction per unit time per unit solid angle about the direction and per unit wavelength about the wavelength, λ; (2) The total radiation intensity, I_b, is used to characterise the amount of energy emitted over the entire wavelength spectrum from $\lambda=0$ to ∞ in a beam and is related to $I_{b\lambda}$ by

$$I_b = \int_0^\infty I_{b\lambda}\,d_\lambda \qquad (4.7)$$

4.8.   Lambert's Cosine Law

This law states that: 'A diffuse surface radiates energy in such a manner that the rate of energy radiated in any particular direction is proportional to the cosine of the angle between the direction under consideration and the normal to the surface'. In fig.4.5 if the rate of energy radiated form surface dA isdQ_n in the direction of normal ON and dQ_ϕ in the direction OP at an angle ϕ to ON and if the surface dA is diffuse, then

$$dQ\phi = dQ_n\cos\phi \qquad\qquad \text{or} \qquad\qquad E_{bn} = E_{bn}\cos\phi \qquad (4.8)$$

Also $\mathrm{Ib}_\phi = \dfrac{1}{\cos_\phi} \dfrac{dE_{b\phi}}{dW} = \dfrac{1}{\cos_\phi} \dfrac{1}{dW}\left(\dfrac{dQ_\phi}{dA}\right)$

$\mathrm{Ib}_\phi = \dfrac{1}{\cos_\phi} \dfrac{dE_{b\phi}}{dW} = \dfrac{1}{\cos_\phi} \dfrac{1}{dW}\left(\dfrac{dQ_\phi}{dA}\right)$

$= \dfrac{1}{\cos_\phi} \dfrac{1}{dW}\left(\dfrac{dQ_n \cos_\phi}{dA}\right) = \dfrac{1}{dW}\left(\dfrac{dQ_n}{dA}\right)$

or $\quad \mathrm{Ib}_\phi = \dfrac{dE_{bn}}{dW} = I_{bn} = I_b = \text{Constant}$ $\qquad\qquad$ (4.9)

This shows that the surfaces which obey Lambet's law, have radiation intensities which are independent of direction. Such surfaces are also called Lambertonian surface.

The radiation intensity and the hemispherical emissive powers of diffuse surfaces have a very simple relationship with each other which can be derived as follows. the solid angle subtended at point O on area dA in direction ϕ at P in (Fig.4.5).

$$dW = \frac{dA_p}{R^2} = \frac{(R\,d\phi)(Rd\Psi \sin\phi)}{R^2} \qquad\qquad (4.10)$$

$$dW = \sin\phi \; d\phi.d\Psi$$

Using Eqn. (4.10) in conjunction with Eqn. (4.6), we get

$$E_b = \int_0^{2\pi} \int_0^{\pi/2} \left(I_{b\phi} \cos_\phi\right)\left(\sin_\phi\right) d\phi\, d\Psi \qquad\qquad (4.11)$$

Putting $\eta = \cos\phi$ in Eqn (4.11), we get

$$E_b = \int_0^{2\pi}\left[\int_0^1 I_{b\phi}\, \eta\; d\eta\right] d\Psi \qquad\qquad (4.12)$$

Equation (4.12) relates E and I_ϕ for a general surface in which I_ϕ is a function of ϕ and hence of η. For diffuse surfaces for which, I_ϕ is independent of ϕ, Eqn (4.12) on integration yields.

$E_b = \pi I_b = \sigma\, T^4$ $\qquad\qquad$ (4.13)

Where $Ib_\phi = Ib = \text{Constant}$. This equation is applicable to diffuse surface only and states that. 'The Total emissive power of a diffuse surface is equal to π times its radiation intensity'

Example 1

Estimate the rate of solar radiation on a plane normal to say rays. Assume the sun to be a black body at a temperature of 5527°C. The diameter of the sun $D = 1.39 \times 10^6$ km and its distance from the earth l is 1.5×10^8 km.

Solution

We know that, $E_b = \pi I_b$

So $\quad I_b = \dfrac{E_b}{\pi} = \dfrac{\sigma T^4}{\pi}$

(The rate of solar radiation per unit area is called density of solar radiation)

Also, $I_b = \dfrac{Q}{Ap} = \dfrac{1}{dW}$

Where dW is the solid angle subtended by the sun at the unit area on the plane.

$$dW = \frac{\pi D^2}{4l^2} = \frac{Pr\,ojected\ area}{l^2}$$

$$\frac{Q}{A\cos\phi} = I_b dW \ (\text{but } \phi \text{ is very small})$$

$$\frac{Q}{A} = \frac{\sigma T^4}{\pi} \cdot \frac{\pi D^2}{4l^2}$$

$$= \frac{\left(0.567 \times 10^{-7}\right) \times \left(5800\right)^4 \left(1.39 \times 10^6\right)^2}{4\left(1.5 \times 10^8\right)^2} = 1377 \text{ W/m}^2$$

4.9.   Laws of Black Body Radiation

We have already discussed that a black body is a perfect absorber and also a perfect emitter, its total intensity and total emissive power are only function of its temperature; it emits energy according to Lambert's Cosine Law. Now let us study the various laws which are obeyed by a black body.

4.9.1. *Planck's Distribution Law*

In 1900 Max Planck (1959) showed by quantum arguments that the spectral distribution of the radiation intensity of a black body is given by the equation.

$$E_{b\lambda}\ \pi I_{b\lambda} = \frac{2\pi C_1}{\lambda^5 \left[\exp\left(C_2 / \lambda T\right) - 1\right]} \tag{4.14}$$

Where λ = wavelength, μ

$\quad$ T = Temperature, K

$C_1 = h\,C_0^2 = 0.595 \times 10^{-8}\ W/m^2$; h being Planck's const.

$\quad$ = 6.625 x 10-34 Js

$C_2 = hC_0/k$ = 1.4387 x10-2mk; k being Boltzmann const.

$\quad$ = 1.38049x10-23 J/mol K

This formula gives the magnitude of the emitted energy at each of the wavelength which comprise the whole radiation spectrum. Equation (4.14) has been plotted in Fig.4.6 where the hemispherical spectrum emissive power is given as a function of wavelength for different temperatures. The curves show the following distinct characteristics of black body radiation: (i) energy emitted at all wavelengths increases with rise in temperature (ii) peak spectral emissive power shifts towards a smaller wavelength at higher temperatures. More precisely, when the temperature of a body is varied, not only does the quantity of emission change but also its distribution over wavelength band. Shown also in Fig.4.6 is the visible spectrum. For a body at 1000K only a small amount of energy is in the visible region. When its temperature is raised further its colour changes from red toward the violet end of the spectrum and at very high temperatures it becomes white, representing radiation composed of a mixture of all the wavelengths. Equation (4.14) is of great importance as it provides quantitative results for the radiation from a black body.

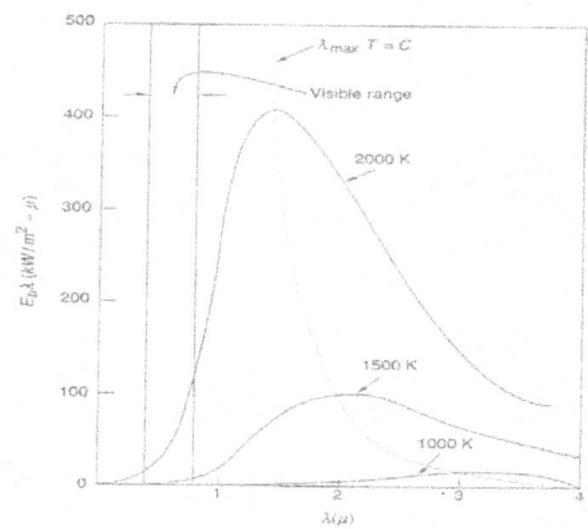

Fig. 4.6: Special Hemispherical Emissive Power for Different Temperatures

Equation (4.14) can be rearranged in the following form:

$$\frac{E_{b\lambda}}{T^5} = \frac{\pi I_{b\lambda}}{T^5} = \frac{2\pi C_1}{(\lambda T)^5 [\exp(C_2 / \lambda T - 1)]} \qquad (4.15)$$

This equation gives the quantity $E_{b\lambda}/T^5$ in terms of the single variable λT. for a given value of λT the ratio of $E_{b\lambda}/T^5$ is the same for all temperatures and their relationship can be presented by a single curve (instead of many in Fig. 4.6) as shown in Fig.4.7 Planck's law given the maximum (black body) radiation intensity at a given wavelength for a given temperature. This intensity serves as a standard with which real surface performance can be compared. There are two approximate forms of Planck's distribution which can be used because of their simplicity. These formulae as given below should be used cautiously due to their limited range of wavelength.

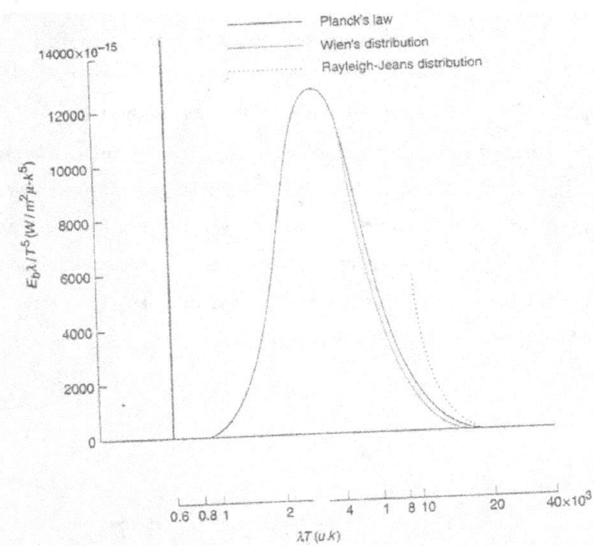

Fig. 4.7: Spectral Distribution of Black Body Hemispherical Emissive Power

4.9.2. *Wien's Formula*

If the term $\exp(C_2/\lambda T)$ is $>> 1$, i.e., for small wavelengths, $\lambda << 1$ Eqn. (4.15) reduces to

$$\frac{I_{b\lambda}}{T^5} = \frac{2C_1}{(\lambda T)^5 \exp(C_2 / \lambda T)} \qquad (4.16)$$

Equation (4.16) is called Wien's formula, and it is accurate within 1 percent for λT less than 3000 (μK).

Rayleigh – Jean's Formula

If λ is large, i.e., for $\lambda >> 1$, $C_2/\lambda T << 1$ and exp $(C_2/\lambda T)$ can be expanded in series to give

$$\exp(C_2/\lambda T) - 1 = 1 + \frac{C_2}{\lambda T} + \frac{1}{2!}\left(\frac{C_2}{\lambda T}\right)^2 + \frac{1}{3!}\left(\frac{C_2}{\lambda T}\right)^3 + \ldots\ldots - 1 \tag{4.17}$$

$$= \frac{C_2}{\lambda T} \text{ for } \lambda T \text{ much larger than } C_2$$

Equation (4.15) then becomes

$$\frac{I_{b\lambda}}{T^5} = \frac{2C_1}{C_2}\frac{1}{(\lambda T)^4} \tag{4.18}$$

This is known as Rayleigh – Jean's Formula. It is accurate within one percent for $(\lambda T) > 8 \times 10^5 \mu K$. Since a black body emits over 99.9 per cent of its energy at λT values below this the formula is well outside the range of thermal radiation. However, the formula is useful for long wave radiation such as radio waves.

Also shown in Fig. 4.7 is a comparison of Wien's and Rayleigh-Jean's approximate formulas with the Planck distribution.

4.9.3. *Wien's Displacement Law*

For a black body emissive spectrum the wavelength λ_{max}, giving the maximum emissive power at a particular temperature, may be found by differentiating Eqn. (4.14) w.r.t. λ and setting it to zero.

$$\lambda_{max} T = \frac{C_2}{5}\frac{1}{1 - \exp(-C_2 / \lambda_{max} T)} \tag{4.19}$$

Which gives $\qquad \lambda_{max} T = C_3 \tag{4.20}$

Which is one form of Wien's displacement law with the value of constant.

$C_3 = 0.289 \times 10^{-2} mK.$

Alternately the value of $\lambda_{max}.$ T can be found at the peak of the distribution curve given in

Fig.4.7 in which we also see that for a given value of λT, the ratio of $\dfrac{E_{b\lambda}}{T^5}$ is the same for all

temperatures.

Another form of Wien's law, therefore, is:

$$\frac{E_{b\lambda max}}{T^5} = \text{const} \quad \text{or} \quad E_{b\lambda max} = C_4 T^5 \tag{4.21}$$

Where,

$C_4 = 1.307 \times 10^{-5}$ W/m^2 K^5.

4.9.4. *Stefan –Boltzmann Law*

It Follows from the definition of total radiation intensity that

$$I_b = \int_0^\infty I_{b\lambda} d\lambda$$

This integral can be evaluated by substituting the value of $I_{b\lambda}$ from Eqn. (4.14)

$$I_b = \int_0^\infty \frac{2C_1}{\lambda^5 [\exp(C_2 / \lambda T) - 1]} d\lambda$$

Now let $C_2/\lambda T = y$

$$\text{Then,} \quad I_b = \frac{2C_1 T^4}{C_2^4} \int_0^\infty \frac{y^3}{e^y - 1} dy \tag{4.22}$$

Which can be evaluated as,

$$\frac{2C_1 T^4}{C_2^4} \int_0^\infty y^3 \left\{ e^{-y} + e^{-2y} + e^{-3y} + ... \right\} dy$$

$$\text{Now} \int_0^\infty y^3 e^{-ny} dy = \frac{3!}{n^{(3+1)}} = \frac{3!}{n^4}$$

$$\therefore \quad I_b = \frac{2C_1 T^4}{C_2^4} \cdot \frac{\pi^4}{15} = \frac{\sigma}{\pi} . T^4 \tag{4.23}$$

$$\text{Where the constant} \quad I_b = \frac{2C_1 T^4}{C_2^4} \left\{ \frac{3!}{1^4} + \frac{3!}{2^4} + \frac{3!}{3^4} + ... \right\} = \frac{2C_1 T^4}{C_2^4} \left(\frac{6\pi^4}{90} \right)$$

$$I_b = \frac{2C_1 T^4}{C_2^4} \cdot \frac{\pi^4}{15} = \frac{\sigma}{\pi} . T^4$$

$$\text{Where the constant} \quad \sigma = \frac{2C_1 T^5}{15 C_2^4} = 0.567 \times 10^{-7} \text{ W} / \text{m}^2 \text{K}^4$$

The total emissive power or hemispherical emissive power of a surface is then.

$$E_b = \int_0^\infty E_{b\lambda}\, d\lambda = \int_0^\infty \pi I_{b\lambda}\, d\lambda = \sigma T^4 \qquad (4.24)$$

Which is known as the Stefan-Boltzmann law and is the same as derived theoretically by Boltzmann and given empirically by Stefan.

Often it is desirable to know how much emission occurs in the wavelength interval between λ_1 and λ_2 at a given surface temperature. This is expressed as fraction of the total emissive power and is designated as $F_{\lambda 1-\lambda 2}$.

$$F\lambda_1 - \lambda_2 = \frac{\displaystyle\int_{\lambda_1}^{\lambda_2} E_{b\lambda}\, d\lambda}{\displaystyle\int_0^\infty E_{b\lambda}\, d\lambda} = \frac{1}{\sigma T^4}\int_{\lambda_1}^{\lambda_2} E_{b\lambda}\, d\lambda \qquad (4.25)$$

$$= \frac{1}{\sigma T^4}\left[\int_0^{\lambda_2} E_{b\lambda}\, d\lambda - \int_0^{\lambda_1} E_{b\lambda}\, d\lambda\right]$$

$$= F_0 - \lambda_2 - F_0 - \lambda_1 \qquad (4.26)$$

Since $E_{b\lambda}$ depends upon T, would require that $F_{0-\lambda}$ be tabulated for each T. we can avoid this complexity by redefining the F function in terms of the only single variable λT. in this manner a universal set of values can be calculated will apply for all temperatures and wavelengths. Rewriting as

$$F\lambda_1-\lambda_2 = F\lambda_1 T - \lambda_2 T$$

$$= \frac{1}{\sigma}\left[\int_0^{\lambda_2 T} \frac{E_{b\lambda}}{T^5} d(\lambda T) - \int_0^{\lambda_1 T} \frac{E_{b\lambda}}{T^5} d(\lambda T)\right]$$

$$= F_0 - \lambda_2 T - F_0 - \lambda_1 T \qquad (4.27)$$

As seen earlier $E_{b\lambda}/T^5$ is only a function of λT so the integrants in Eqn. (4.27) are dependent only on λT. The following examples illustrate the use of the $F_{0-\lambda T}$ function.

Example 4.2

A black body is kept at a temperature of 727°C. Estimate the fraction of the thermal radiation emitted by the surface in the wavelength band 1.0 and 5.0μ.

Solution

For T = 727°C = 727+273 = 1000K, we have

$\lambda_1 T = 1.0 \times 1000 = 1000 \mu K.$

4.10. Radiation Exchange between Surfaces

Radiant energy exchange between surfaces depends not only on the emission, absorption and reflection characteristics of the surfaces but also on their geometrical arrangement.

This heat eschange will be affected further due to the presence of partially emitting and absorbing medium in between the surfaces. To account this radiation exchange, following assumptions are made.

1. All surfaces are considered to be either black or gray.
2. Radiation and reflection process are assumed to be diffuse.
3. The absorptivity of a surface is taken equal to its emissivity and independent of temperature of the source of the incident radiation.

Radiation Exchange between Two Black Surfaces Separated by a Non Absorbing Medium

Let us consider two black bodies separated by a non absorbing medium. The problem is to determine the net radiation heat exchange between them.

Consider area elements dA_1 and dA_2 on the two surfaces. The distance between them is r and the angles, the normal to the two area elements make with the line joining them are θ_1 and θ_2.

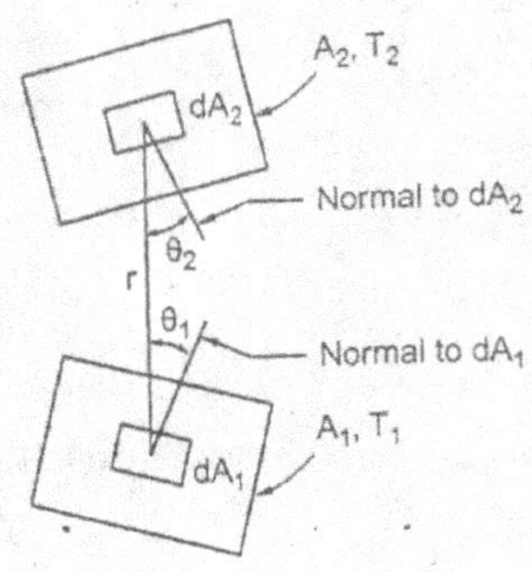

Fig. 4.8: Radiation Heat Exchange between Two Black Surfaces

Fig.4.8 shows the projection of d A_1 normal to the between the centres. The projected area is d $A_1 \cos\theta_1$. Energy leaving d A_1 in direction θ_1.

$$= 1_{n1} \, dA_1 \cos\theta_1 \tag{4.28}$$

where1_{n1} – intensity of radiation at surface A_1 we know that

Intensity of radiation, $1_{n1} = \dfrac{E_{b1}}{\pi}$

Radiation arriving at any normal to r will depend on the solid angle subtended by it.

Let $d\omega_1$ be subtended at dA_1 by dA_2 and $d\omega_2$ subtended dA_2 by dA_1.

$$\text{so, } d\omega_1 = \frac{dA_2 \cos\theta_2}{r^2} \tag{4.29}$$

$$d\omega_2 = \frac{dA_1 \cos\theta_1}{r^2} \tag{4.30}$$

The rate of radiant energy leaving dA_1 and striking on dA_1 given by

$dQ_{1\text{-}2} = 1_{n1}\, dA_1 \cos\theta_1 \times d\omega_1$

$$= 1_{n1}\, dA_1 \cos\theta_1 \times \frac{dA_2 \cos\theta_2}{r^2}$$

$$\boxed{dQ_{1-2} = \frac{1n_1 \cos\theta_2 \cos\theta_2\, dA_1\, dA_2}{r^2}} \tag{4.31}$$

The rate energy radiated by dA_2 and absorbed by dA_2 given by

$dQ_{2\text{-}1} = 1_{n2}\, dA_2 \cos\theta_2 \times d\omega_2$

$$dQ_{2\text{-}1} = 1_{n2}\, dA_2 \cos\theta_2 \times \frac{dA_1 \cos\theta_1}{r^2}$$

$$\boxed{dQ_2 = \frac{1_{n_2} \cos\theta_1 \cos\theta_2\, dA_1\, dA_2}{r^2}} \tag{4.32}$$

The net rate of heat transfer between dA_1 and A_2 is

$dQ_{12} = dQ_{1\text{-}2} - dQ_{2\text{-}1}$

$$= \frac{1_{n_1} \cos\theta_1 \cos\theta_2\, dA_1\, dA_2}{r^2} - \frac{1_{n_2} \cos\theta_1 \cos\theta_2\, dA_1\, dA_2}{r^2}$$

$$dQ_{12} = (1_{n1} - 1_{n2}) = \left[\frac{\cos\theta_1 \cos\theta_2\, dA_1\, dA_2}{\pi r^2} \right]$$

show

$$1\pi = \frac{E_\delta}{\pi}$$

$$dQ_{12} = (E_{\delta 1} - E_{\delta 2}) \left[\frac{\cos\theta_1 \cos\theta_2 \, dA_1 \, dA_2}{\pi r^2} \right]$$

From Stefan Boltzmann law, we know

$$E_b = \sigma T^4$$

$$dQ_{12} = \left(\sigma T_1^4 - \sigma T_2^4\right) \left[\frac{\cos\theta_1 \cos\theta_2 \, dA_1 \, dA_2}{\pi r^2} \right]$$

$$dQ_{12} = \left(T_1^4 - T_2^4\right) \left[\frac{\sigma \cos\theta_1 \cos\theta_2 \, dA_1 \, dA_2}{\pi r^2} \right] \tag{4.33}$$

The rate of total net heat transfer for the total areas A_1 and A_2

$$\int dQ_{12}$$

$$= \int\limits_{A1} \int\limits_{A2} \left(T_1^4 - T_2^4\right) \left[\frac{\sigma \cos\theta_1 \cos Q_2 \, dA_1 \, dA_2}{\pi r^2} \right] \tag{4.34}$$

$$Q_{12} = \sigma\left(T_1^4 - T_2^4\right) \int\limits_{A1} \int\limits_{A2} \left[\frac{\cos\theta_1 \cos Q_2 \, dA_1 \, dA_2}{\pi r^2} \right]$$

From equation (4.31)

$$dQ_{1\text{-}2} = \frac{1_{nl} \cos\theta_1 \cos Q_2 \, dA_1 \, dA_2}{r^2}$$

$$\int dQ_{1-2} = 1_{nl} \int\limits_{A1} \int\limits_{A2} \frac{\cos Q_1 \cos Q_2 \, dA_1 \, dA_2}{r^2}$$

$$\frac{Q_{1-2}}{Q_1} = F_{12}$$

$$\frac{1}{A_1} \int\limits_{A1} \int\limits_{A2} \frac{\cos\theta_1 \cos\theta_2 \, dA_1 \, dA_2}{\pi r^2}$$

$$Q_{1-2} = \frac{\sigma\left(T_1^4\right)}{\pi} \int\limits_{A1} \int\limits_{A2} \frac{\cos Q_1 \cos Q_2 \, dA_1 \, dA_2}{r^2} \tag{4.35}$$

Total energy radiated by A_1 is given by

$$Q_1 = A_1 \, \sigma \, T_1^4$$

$$\Rightarrow \frac{Q_{1-2}}{Q_1} = \frac{\sigma T_4^4 \int\limits_{A1} \int\limits_{A2} \dfrac{\cos Q_1 \cos Q_2 \, dA_1 \, dA_2}{\pi r^2}}{A1 \, \sigma \, T_1^4}$$

$$\frac{Q_{1-2}}{Q_1} = \frac{1}{A_1} \int\limits_{A1}\int\limits_{A2} \frac{\cos\theta_1 \cos\theta_2 \ dA_1 \ dA_2}{\pi r^2} \qquad (4.36)$$

$$\frac{Q_{1-2}}{Q_1} = F_{12} \qquad (4.37)$$

Where F_{12} = shape factor (0r) configuration factor (or) view factor. Shape factor is defined as "The fraction of radiative energy that is diffused from one surface and strikes the other surface directly with no intervening reflections"

$$Q_{1\text{-}2} = F_{12} \times Q_1$$

$$\boxed{Q_{2\text{-}1} = F_{1\text{-}2}\,A_1} \qquad (4.38)$$

Similarly,

$$Q_{2\text{-}1} = \sigma_{2-1} = T_2^4 \int\limits_{A1}\int\limits_{A2} \frac{\cos\theta_1 \cos\theta_2 \ dA_1 \ dA_2}{\pi r^2} \qquad (4.39)$$

The total energy radiated by A_2 is given by

$$Q_2 = A_2\,\alpha\,T_2^4$$

$$\frac{Q_{2-1}}{Q_2} = \frac{1}{A_2} \int\limits_{A1}\int\limits_{A2} \frac{\cos\theta_1 \cos\theta_2 \ dA_1 \ dA_2}{\pi r^2}$$

$$\frac{Q_{2-1}}{Q_2} = F_{2-1}$$

Where F_{21}- Shape factor of A_2 with respect to A_1

$$Q_{2\text{-}1} = F_{2\text{-}1}\,Q_2$$

$$\boxed{Q_{2\text{-}1} = F_{2\text{-}1}A_1\ \sigma\,T_2^4} \qquad (4.40)$$

From equation (4.18), (4.20), we know that, $A_1\,F_{1\text{-}2} - A_2\,F_{2\text{-}1}$

This is known as reciprocity theorem. Thus the net rate of heat transfer between two surface A_1 and A_2 is given by

This equation is applicable to black surfaces only. If surface having

Emissivity –

$$\boxed{\begin{array}{c} Q_{12} = A_1 F_{12}\,\sigma\left[1 - \dfrac{\sigma_1(1-\sigma_2)}{1-(1-\sigma_1)(1-\sigma_2)}\right] \\[3mm] Q_{12} = \overline{\varepsilon}\,A_1\,F_{12}\,\sigma\left(T_1^4 - T_2^4\right) F_{12}\,\sigma\left(T_1^4 - T_2^4\right) \end{array}}$$

$\overline{\varepsilon}$ - Emissivity of surface

4.11. Shape Factor

Shape factor is defined as "The fraction of radiative energy that is diffused from onn surface and strikes the other surfine directly with no intervening reflections".

4.12. Shape Factor Algebra (or) View Factor Algebra

In order to compute the shape factor for certain geomet arrangements for which shape factors charts or equations are available, the concept of shape factor as a fraction of intercepted energy and the reciprocity theorem can be used.

The shape factors for these geometries can be derived in term of known shape factors of other geometries. The interrelation between various factors is called shape factor algebra.

4.13. Heat Exchange between Two Non-Black (Gray) Parallel Planes

Consider two very large parallel gray Surfaces of areas A1 and at a small distance apart, and exchanging radiation as shown in fig.4.9.

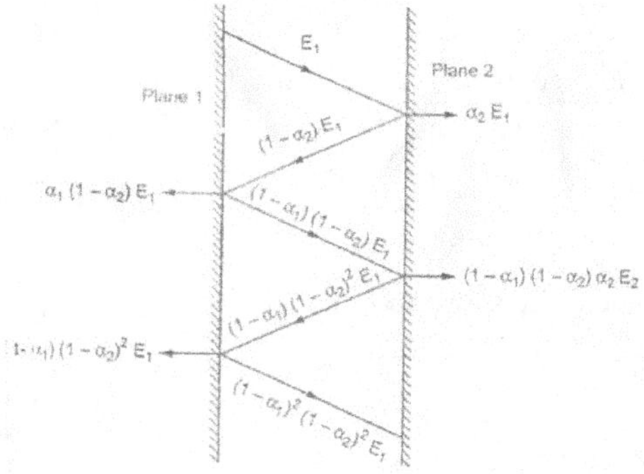

Fig 4.9

i.e., $T_1 \cdot \alpha_1$ and ε_1 be the temperature, absorptivity and absorptivity of the surface 1.

Similarly T_2, α_2 and ε_2 be the temperature, absorptivity and absorptivity of the surface 2.

The following assumptions are made for the analysis.

The configuration factor of either surface is unity.

There is no absorbing medium in between the surfaces.

The emissive and reflective properties are constant for over all surfaces.

The surface 1 emits radiant energy E_1 which falls on the surface 2. Out of this, a part of $\alpha_2 E_1$ is absorbed by the surface 2 and the remainder $(1-\alpha_2) E_1$ is reflected back to surface 1.

On reaching surface 1. a part α_1 $(1-\alpha_2)$ E_1 is absorbed and the remainder $(1-\alpha_1)$ $(1-\alpha_2)$ E_1 is reflected. This process will go on continuing.

The rate of radiant energy leaving surface 1is given by

$$Q_1 = E_1 - [\alpha_1 (1-\alpha_2) E_1 + \alpha_1 (1-\alpha_1)^2 E_1 + \alpha_1 (1-\alpha_2)^2 (1-\alpha_2)^3 E_1 + \ldots\ldots]$$

$$= E_1 - \alpha_1 (1-\alpha_2) E_1 [1+(1-\alpha_1)(1-\alpha_2) + (1-\alpha_2)^2 (1-\alpha_2)^2 + \ldots\ldots]$$

$$Q_1 = E_1 - \alpha_1 (1-\alpha_2) E_1 [1 + P + P^2 + \ldots\ldots]\ldots \tag{4.41}$$

Where $P = (1-\alpha_1)(1-\alpha_2)$

Since α_1 and α_2 are less than unity, P will be less than unity. As P< 1, the series $1+P+P^2 + \ldots$

When the infinity extended gives $\dfrac{1}{1-P}$

$$(4.41) \Rightarrow Q_1 = E_1 - \alpha_1(1-\alpha_2) E_1 \times \dfrac{1}{1-P}$$

$$= E_1 - \dfrac{\sigma_1(1-\sigma_2)E_1}{1-P}$$

Substitute P value

$$Q_1 = E_1 \left[1 - \dfrac{\sigma_1(1-\sigma_2)}{1-(1-\sigma_1)(1-\sigma_2)} \right]$$

From Kirchhoff's law, we know that, emissivity and absorptivity of asurface are equal.

$$\Rightarrow \alpha_1 = \varepsilon_1 \text{ and } \alpha_2 = \varepsilon_2$$

$$Q_1 = E_1 \left[1 - \dfrac{\varepsilon_1(1-\varepsilon_2)}{1-(1-\varepsilon_1)(1-\varepsilon_2)} \right]$$

$$= E_1 \left[\dfrac{1-(1-\varepsilon_1)(1-\varepsilon_2)-\varepsilon_1(1-\varepsilon_2)}{1-(1-\varepsilon_1)(1-\varepsilon_2)} \right]$$

$$= E_1 \left[\dfrac{1-[1-\varepsilon_2-\varepsilon_1+\varepsilon_1\varepsilon_2]-\varepsilon_1(1-\varepsilon_2)}{1-(1-\varepsilon_2-\varepsilon_1+\varepsilon_1\varepsilon_2)} \right]$$

$$= E_1 \left[\dfrac{1-1+\varepsilon_2+\varepsilon_1-\varepsilon_1\varepsilon_2-\varepsilon_1+\varepsilon_1\varepsilon_2}{1-1+\varepsilon_2+\varepsilon_1-\varepsilon_1\varepsilon_2} \right]$$

$$Q_{12} = \dfrac{E_1\varepsilon_2 - E_2\varepsilon_1}{\varepsilon_1+\varepsilon_2-\varepsilon_1\varepsilon_2}$$

$$\boxed{Q_1 = \dfrac{E1\,\varepsilon 2}{\varepsilon_1+\varepsilon_2-\varepsilon_1\varepsilon_2}}$$

Similarly,

The rate of radiant energy leaving surface 2 is given by

$$\boxed{\frac{E_1\ \varepsilon_2}{\varepsilon_1+\varepsilon_2-\varepsilon_1\ \varepsilon_2}-\frac{E_2\ \varepsilon_2}{\varepsilon_1+\varepsilon_2-\varepsilon_1\ \varepsilon_2}}$$

The net radiative heat exchange from surface 1 to 2 is given by

$Q_{12} = Q_1 - Q_2$

$$= \frac{E_1\ \varepsilon_2}{\varepsilon_1+\varepsilon_2-\varepsilon_1\ \varepsilon_2}-\frac{E_2\ \varepsilon_2}{\varepsilon_1+\varepsilon_2-\varepsilon_1\ \varepsilon_2}$$

$$Q_{12} = \frac{E_1\ \varepsilon_2 - E_2\ \varepsilon_1}{\varepsilon_1+\varepsilon_2-\varepsilon_1\ \varepsilon_2}$$

From Stefan –Boltzmann law, we know that,

$E_b = \alpha\ T^4$

$\Rightarrow\quad E_1 = \alpha\ T_1^4$

$$\boxed{\begin{aligned} E_1 &= \varepsilon_1\ \alpha\ T_1^4 \\[2mm] E_2 &= \varepsilon_2\ \alpha\ T_2^4 \end{aligned}}$$

Substitute E_1 = and E_2 values in equation (4.26).

$$Q_2 = \overline{\varepsilon}\ \alpha\ A\ [T_1^4 - T_2^4]$$

$$= \frac{\varepsilon_1\ \varepsilon_2\ \alpha\ [T_1^4 - T_2^4]}{\varepsilon_1+\varepsilon_2-\varepsilon_1\ \varepsilon_2}$$

$$= \frac{\varepsilon_1\ \varepsilon_2}{\varepsilon_1+\varepsilon_2-\varepsilon_1\ \varepsilon_2}\ \text{x}\ \alpha\ [T_1^4 - T_2^4]$$

$$Q_{12} = \overline{\varepsilon}\ \alpha\ [T_1^4 - T_2^4] \tag{4.42}$$

Where, $\overline{\varepsilon}\ = \dfrac{\varepsilon_1\ \varepsilon_2}{\varepsilon_1+\varepsilon_2-\varepsilon_1\ \varepsilon_2}$

Divide by $\varepsilon_1, \varepsilon_2$.

$$\Rightarrow \overline{\varepsilon}\ = \frac{\dfrac{\varepsilon_1\ \varepsilon_2}{\varepsilon_1\ \varepsilon_2}}{\dfrac{\varepsilon_1+\varepsilon_2-\varepsilon_1\ \varepsilon_2}{\varepsilon_1\ \varepsilon_2}}$$

$$\bar{\varepsilon} \;=\; \cfrac{1}{\cfrac{1}{\varepsilon_2}+\cfrac{1}{\varepsilon_1}-1}$$

$$\boxed{\;\bar{\varepsilon} \;=\; \cfrac{1}{\cfrac{1}{\varepsilon_1}+\cfrac{1}{\varepsilon_2}-1}\;}$$

Heat exchange between two parallel surface is given by considering Area).

$$Q_2 = \bar{\varepsilon}\,\alpha\,A\,[T_1^4 - T_2^4] \tag{4.43}$$

[From equation (4.42)]

$$\text{Where}\quad \bar{\varepsilon} \;=\; \cfrac{1}{\cfrac{1}{\varepsilon_1}+\cfrac{1}{\varepsilon_2}-1}$$

4.14. Heat Exchange between Two Large Concentric Cylinders Or Spheres

Consider two large concentric cylinders of areas A_1 and A_2 exchanging radiation as shown in fig. 4.10.

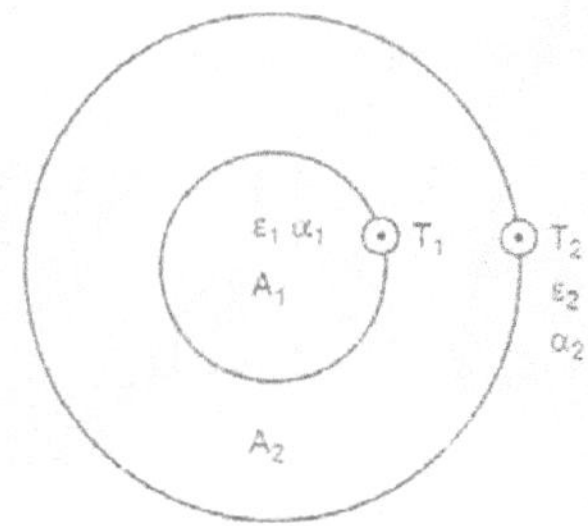

Fig. 4.10

Let T_1, α_1 and ε_1 be the temperature, absorptivity and emissivity of the Inner cylinder.

Similarly T_2, α_2 and ε_2 be the temperature absorptivity and emissivity of the outer cylinder.

We can use the technique as we have used in parallel plates accept.

$$F_{12}\,A_1 = F_{21}\,A_2$$

But $\qquad F_{12} = 1$

$$\Rightarrow \qquad F_{21} = \frac{A_1}{A_2}$$

Considering the energy emitted by inner cylinder,

1. Inner cylinder emits the energy $= E_1$

2. Outer cylinder absorbs energy $= \alpha_2 E_1 = \varepsilon_2 E_1$ $\qquad\qquad [\because \alpha_2 = \varepsilon_2]$

3. Outer Cylinder reflects energy $= E_1(1-\varepsilon_2)$

4. Inner cylinder absorbs energy $= E_1(1-\varepsilon_2)\,F_{21}\alpha_1 = E_1(1-\varepsilon_2)\dfrac{A_1}{A_2}\varepsilon_1$

$$\left[\because F_{21} = \dfrac{A_1}{A_2},\ \alpha_1 = \varepsilon_1 \right]$$

5. Inner cylinder reflects energy
$$= E_1(1-\varepsilon_2) - E_1(1-\varepsilon_2)\varepsilon_1\dfrac{A_1}{A_2}$$

$$= E_1(1-\varepsilon_2)\left[1-\varepsilon_1\dfrac{A_1}{A_2}\right]$$

6. Energy absorbed by the inner cylinder on the second reflection

$$= E_1(1-\varepsilon_2)^2\varepsilon_1\dfrac{A_1}{A_2}\left[1-\dfrac{A_1}{A_2}\varepsilon_1\right]$$

This absorption and reflection go on indefinitely. So we can find the net energy lost by the inner cylinder considering infinite times absorption and reflections.

Heat lost by the inner cylinder per unit area is given by

$$Q_1 = Q_1 = E_1 - E_1(1-\varepsilon_2)\varepsilon_1\dfrac{A_1}{A_2} - E_1(1-\varepsilon_2)^2\varepsilon_1\dfrac{A_1}{A_2}\left[1-\dfrac{A_1}{A_2}\varepsilon_1\right] +$$

$$= E_1\left[1-\dfrac{A_1}{A_2}\varepsilon_1(1-\varepsilon_2)-(1-\varepsilon_2)^2\varepsilon_1\dfrac{A_1}{A_2}\left\{1-\dfrac{A_1}{A_2}\varepsilon_1\right\}+ \ldots\ldots\ldots\ldots\right]$$

$$= E_1\left[1-\dfrac{A_1}{A_2}\varepsilon_1(1-\varepsilon_2)\left\{1+(1-\varepsilon_2)\left(1-\dfrac{A_1}{A_2}\varepsilon_1\right)+(1-\varepsilon_2)^2 \times \left(1-\dfrac{A_1}{A_2}\varepsilon_1\right)^2 + \ldots\ldots\ldots\right\}\right]$$

$$= E_1\left[\dfrac{\dfrac{A_1}{A_2}\varepsilon_1(1-\varepsilon_2)}{1-(1-\varepsilon_2)\left(1-\dfrac{A_1}{A_2}\varepsilon_1\right)}\right]$$

$$Q_1 = \left[\dfrac{E_1\varepsilon_2}{\dfrac{A_1}{A_2}\varepsilon_1 + \varepsilon_2 - \dfrac{A_1}{A_2}\varepsilon_1\varepsilon_2}\right]$$

$$Q_1 = \left[\frac{E_1\varepsilon_2}{\dfrac{A_1}{A_2}\varepsilon_1 + \varepsilon_2 - \dfrac{A_1}{A_2}\varepsilon_1\varepsilon_2} \right]$$

Similarly, Heat lost by the outer cylinder per unit area given by

$$Q_2 = \left[\frac{E_1\varepsilon_2 \dfrac{A_1}{A_2}}{\dfrac{A_1}{A_2}\varepsilon_1 + \varepsilon_2 - \dfrac{A_1}{A_2}\varepsilon_1\varepsilon_2} \right]$$

The net radiation heat transfer between the inner and outer concentric cylinders is given by $Q_{12} = Q_1 - Q_2$

$$Q_{12} = \frac{E_1\varepsilon_2}{\dfrac{A_1}{A_2}\varepsilon_1 + \varepsilon_2 - \dfrac{A_1}{A_2}\varepsilon_1\varepsilon_2} - \frac{E_2\varepsilon_1 \dfrac{A_1}{A_2}}{\dfrac{A_1}{A_2}\varepsilon_1 + \varepsilon_2 - \dfrac{A_1}{A_2}\varepsilon_1\varepsilon_2}$$

Considering area A_1 and A_2

$$\Rightarrow \quad Q_{12} = \frac{A_1 E_1\varepsilon_2}{\dfrac{A_1}{A_2}\varepsilon_1 + \varepsilon_2 - \dfrac{A_1}{A_2}\varepsilon_1\varepsilon_2} - \frac{A_2 E_2\varepsilon_1 \dfrac{A_1}{A_2}}{\dfrac{A_1}{A_2}\varepsilon_1 + \varepsilon_2 - \dfrac{A_1}{A_2}\varepsilon_1\varepsilon_2}$$

$$\Rightarrow \quad Q_{12} = \frac{A_1 E_1\varepsilon_2 - A_2 E_2\varepsilon_1}{\dfrac{A_1}{A_2}\varepsilon_1 + \varepsilon_2 - \dfrac{A_1}{A_2}\varepsilon_1\varepsilon_2} \tag{4.44}$$

From Stefan – Boltzmann law, we know that,

$$E_b \quad = \quad \varepsilon\sigma\, T^4$$

$$\Rightarrow \quad E_1 \quad = \quad \varepsilon_1\, \sigma\, T_1^4$$

$$\Rightarrow \quad E_2 \quad = \quad \varepsilon_2\, \sigma\, T_2^4$$

Substituting E_1 and E_2 values in the equation (4.44),

$$\Rightarrow \quad Q_{12} = \frac{A_1\varepsilon_1 T_1^4\,\varepsilon_2 - A_1\varepsilon_2\sigma T_2^4\varepsilon_1}{\dfrac{A_1}{A_2}\varepsilon_1 + \varepsilon_2 - \dfrac{A_1}{A_2}\varepsilon_1\varepsilon_2}$$

$$= \frac{A_1\sigma\,\varepsilon_1\varepsilon_2\,[T_1^4 - T_2^4]}{\left[\dfrac{A_1}{A_2}\varepsilon_1\varepsilon_2 \left(\dfrac{1}{\varepsilon_2} - 1 \right) \right] + \varepsilon_2}$$

Dividing by ε_1, ε_2,

$$\Rightarrow \quad Q_{12} = \frac{A_1 \sigma [T_1^4 - T_2^4]}{\dfrac{A_1}{A_2}\left(\dfrac{1}{\varepsilon_2} - 1\right) + \dfrac{1}{\varepsilon_1}}$$

$$Q_{12} = \frac{A_1 \sigma [T_1^4 - T_2^4]}{\dfrac{1}{\varepsilon_1} + \dfrac{A_1}{A_2}\left(\dfrac{1}{\varepsilon_2} - 1\right)}$$

$$Q_{12} = \overline{\varepsilon}\, A_1 \sigma (T_1^4 - T_2^4)$$

Where

$$\overline{\varepsilon} = \frac{1}{\dfrac{1}{\varepsilon_1} + \dfrac{A_1}{A_2}\left(\dfrac{1}{\varepsilon_2} - 1\right)}$$

For Cylinders, Area, $A = 2\pi r L$

For Sphere, Area, $A = 4\pi r^2$

4.15. Radiation Shield

Radiation shields constructed from low emissively (high reflective) materials. It is used to reduce the net radiation transfer between two surfaces. Let us consider two parallel planes 1 and 2 each of area A at temperatures T_1 and T_2 respectively. A radiation shield is placed in between them as shown in Fig 4.11.

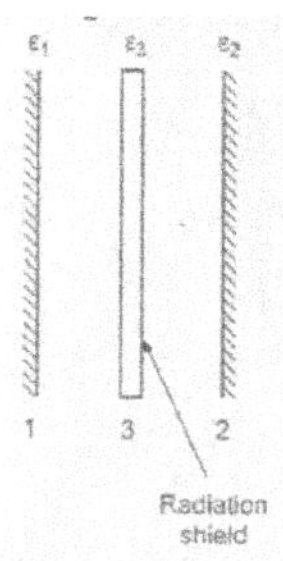

Fig. 4.11: Radiation Shield

The net heat exchange between parallel plates without shield is given by

$$Q_{12} = \frac{A\sigma(T_1^4 - T_2^4)}{\dfrac{1}{\varepsilon_1} + \dfrac{1}{\varepsilon_2} - 1} \qquad (4.45)$$

[From equation No. (4.42)]

Where

A   - Area in m².

σ - Stefan Boltzmann constant = 5.67 x 10⁻⁸ W/m² K⁴

$\varepsilon_1, \varepsilon_2$ - Emissivities of surface 1 and 2 respectively.

T_1, T_2 - Temperature of surface 1 and 2 respectively.

Heat Exchange between 1 and 3 is

$$Q_{13} = \frac{A\sigma\left(T_1^4 - T_3^4\right)}{\dfrac{1}{\varepsilon_1} + \dfrac{1}{\varepsilon_3} - 1} \qquad (4.46)$$

Heat exchange between 3 and 2 is

$$Q_{32} = \frac{A\sigma(T_3^4 - T_2^4)}{\dfrac{1}{\varepsilon_1} + \dfrac{1}{\varepsilon_2} - 1} \qquad (4.47)$$

From equation (4.46),

$$T_1^4 - T_3^4 = \frac{Q_{13}\left(\dfrac{1}{\varepsilon_1} + \dfrac{1}{\varepsilon_3} - 1\right)}{A\sigma}$$

$$\Rightarrow \quad T_3^4 = T_1^4 - \frac{Q_{13}\left(\dfrac{1}{\varepsilon_1} + \dfrac{1}{\varepsilon_3} - 1\right)}{A\sigma}$$

Substitute T_3^4 Value in equation (4.47)

$$(4.47) \Rightarrow \quad Q_{32} = \frac{A\sigma\left[\left(T_1^4 - \dfrac{Q_{13}\left(\dfrac{1}{\varepsilon_1} + \dfrac{1}{\varepsilon_3} - 1\right)}{A\sigma}\right) - T_2^4\right]}{\dfrac{1}{\varepsilon_3} + \dfrac{1}{\varepsilon_2} - 1}$$

$$\Rightarrow \quad Q_{12}\left(\frac{1}{\varepsilon_3} + \frac{1}{\varepsilon_2} - 1\right) = A\sigma T_1^4 - Q_{13}\left(\frac{1}{\varepsilon_1} + \frac{1}{\varepsilon_3} - 1\right) - A\sigma$$

$$\Rightarrow \quad Q_{12}\left(\frac{1}{\varepsilon_3} + \frac{1}{\varepsilon_2} - 1\right) + Q_{13}\left(\frac{1}{\varepsilon_1} + \frac{1}{\varepsilon_3} - 1\right) = A\sigma T_1^4 - A\sigma T_2^4$$

Under equilibrium condition

$$Q_{13} = Q_{32}$$

$$\Rightarrow \quad Q_{13}\left[\left(\frac{1}{\varepsilon_3}+\frac{1}{\varepsilon_2}-1\right)+\left(\frac{1}{\varepsilon_1}+\frac{1}{\varepsilon_3}-1\right)\right] = A\sigma(T_1^4 - T_2^4)$$

$$\Rightarrow \quad Q_{13} = \frac{A\sigma(T_1^4 - T_2^4)}{\left(\dfrac{1}{\varepsilon_3}+\dfrac{1}{\varepsilon_2}-1\right)+\left(\dfrac{1}{\varepsilon_1}+\dfrac{1}{\varepsilon_3}-1\right)} \tag{4.48}$$

Dividing the equation (4.37) by equation (4.34),

$$\frac{Q_{13}}{Q_{12}} = \frac{\left(\dfrac{1}{\varepsilon_1}+\dfrac{1}{\varepsilon_2}-1\right)}{\left(\dfrac{1}{\varepsilon_3}+\dfrac{1}{\varepsilon_2}-1\right)+\left(\dfrac{1}{\varepsilon_1}+\dfrac{1}{\varepsilon_3}-1\right)}$$

If $\quad \varepsilon_1 = \varepsilon_2 = \varepsilon_3$

$$\frac{Q_{13}}{Q_{12}} = \frac{1}{2}$$

$$Q_{13} = \frac{1}{2}Q_{12} \text{ (or) } Q_{32} = \frac{1}{2}Q_{12}$$

Thus by inserting one shield between two parallel surfaces, the direct radiation heat transfer between them is halved.

4.16. Electrical Network Analogy for Thermal Radiation Systems

An alternative approach for analysing thermal radiation between gray or black surfaces is called electrical network analogy. This approach is more direct, more general and much simpler than the one presented in the preceding sections. The two terms often used in the electrical analogy approach the irrational and radiosity.

Irradiation.G, is the total radiation incident upon a surface per unit time and per unit area. τ per unit area. Both these quantities may be thought of as fluxes, and are expressed in W/m². The radiosity consists of two parts' that which is reflected by the surface and that which is emitted by the surface. The first part is equal ρG to and the second part is the emissive power the surfaces. Then

$$J = \rho G + \varepsilon\, E_b \tag{4.49}$$

Assuming the surface to be opaque ($\tau = 0$)

$$\rho = 1 - \alpha = 1 - \varepsilon$$

so that $\qquad J = (1 - \varepsilon)G + \varepsilon E_b \qquad (4.50)$

or $\quad G \quad = \quad (J - \varepsilon E_b)/(1 - \varepsilon) \qquad (4.51)$

The net energy leaving a surface is the difference between its raiosity and irradiation.

$$\frac{Q}{A} = J - G$$

$$\frac{Q}{A} = J - \left(\frac{J - \varepsilon E_b}{1 - \varepsilon} \right) \qquad (4.52)$$

$$Q = \left(\frac{\varepsilon A}{1 - \varepsilon} \right)(E_b - J) = \frac{E_b - J}{(1 - \varepsilon)/\varepsilon A} \qquad (4.53)$$

If we examine Eqn. (4.53) closely we at once realize that because of the nature of the surface not all the energy which could be emitted or absorbed is actually emitted or absorbed. In other words, there is some kind of restriction of the flow of radiant energy, which may well be called a surface resistance. The concept leads us to formulate some kind of analogy with Lhm's law relating current, I, and potential difference. ΔV with resistance, R

$$I = \frac{\Delta V}{R}$$

If we consider the net heat flow rate as a current, and $(E_b\text{-}J)$ as the potential difference, them a comparison of Eqn. (4.53) with Ohm's law indicates that

$$\text{Surface resistance} = \frac{1 - \varepsilon}{\varepsilon A} \qquad (4.54)$$

Now consider the exchange of radiant energy between two surfaces. A_1 and A_2. We know that because of the orientation of the surfaces, all the area available for radiation from one surface is not covered by the second surface. This is taken care of by the shape factor and area. So again there is restriction on the free flow of energy between two surfaces because of their orientation (shape) and this restriction may be termed shape resistance. The shape resistance can be found as follows:

Of the total radiation which leaves surface 1, the amount that reaches surface 2 is

$Q = AF_{12}J_1$

Similarly for surface 2

$Q_2 = A_2 F_{21}J_2$

The net interchange from surface 1 to 2 is

$Q_{12}=AF_{12}J_1=A_2F_{21}J_2$

Since $A_1F_{12}=A_2F_{21}$

$Q_{12}=AqF_{12}(J_1-J_2)$

$$= \frac{J_1 - J_2}{1/A_2F_{12}} \tag{4.55}$$

A comparison with Ohm's indicates that

$$\text{Shape resistance} = \frac{1}{AF_{12}} \tag{4.56}$$

We may this say that two surfaces which exchange hear may each be considered as having a surface resistance $(1\text{-}\varepsilon)\varepsilon A$ and a shape resistance $\dfrac{1}{AF_{12}}$ between their radiosity potentials.

The radiation network for this system is shown in Fig. 10.11. In this case the net heat exchange would be the overall potential difference divided by the sum of the resistances.

$$Q_{12} = \frac{Eb_1 - Eb_2}{(1-\epsilon_1)/\epsilon_1 A_1 + 1/A_1F_{12} + (1-\epsilon_2)/\epsilon_2 A_2}$$

$$= \frac{\sigma(T^4_1 - T^4_1)}{(1-\epsilon_1)/\epsilon_1 A_1 + F_{12} + (1-\epsilon_2)\epsilon_2 A_2)} \tag{4.57}$$

Fig. 4.12: Radiation Network for Two Gray Surfaces

The radiation network of Fig 4.12 and the energy exchange equation. Eqn 4.57 are for any two gray surface exchanging radiation only between themselves. We shall now demonstrate how the network method can be employed by taking some special cases.

(i) Radiation Exchange between Two Black Surfaces

As the black surfaces are perfect emitters and absorbers, so their reflectivity, $\rho=0$ or $(1-\epsilon)=0$, and thus $J = E_b$, A black surface offers so surface resistance, and therefore, the radiation network between two black surfaces will comprise only one shape resistance as shown in Fig.4.13 From Eqn,. (4.57), for black surfaces,

$$Q_{12} = \frac{\sigma(T^4_{\;1} - T^4_{\;2})}{1/A_1F_{12}} = A_1 F_{12} \sigma(T^4_1 - T^4_2)$$

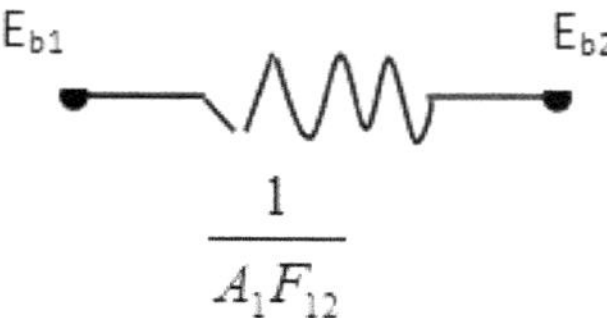

Fig. 4.13: Radiation Network for Two Black Surfaces

(ii) Radiant Exchange between Two Black Surfaces Connected by a Third Refractory Surface

The network shown in Fig.4.14 represents a system of two black surfaces 1 and 2 exchanging radiation via a third reradiating surface. As a reradiating surface exchanges heat with other surfaces, its emissive power E_{bR}, floats between E_{b1} and E_{b2} at a level which depends upon the magnitudes of Eb_1 and Eb_2 and the shape resistances between E_{bR}, and E_{b1}, E_{b2}. In other words the system may be considered as a series-parallel circuit as shown in Fig.4.14.

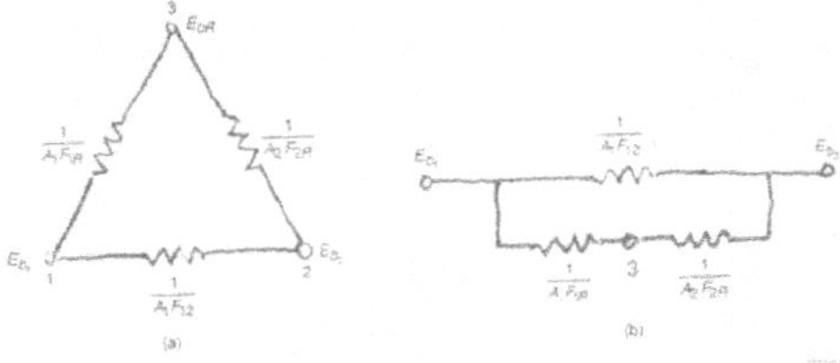

Fig. 4.14: Radiation Network for Two Black Surfaces Connected by a Third Reradiating Surfaces

The total resistance of the circuit is

$$\frac{1}{R} = \frac{1}{1/A_1F_{12}} + \frac{1}{1/A_1F_{1R} + 1/A_2F_{2R}}$$

$$Q_{12} = \frac{E_{b1} - E_{b2}}{R} = (E_{b1} - E_{b2})\left[A_1F_{12} + \frac{1}{1/A_1F_{1R} + 1/A_2F_{2R}} \right]$$

$$= \left[A_1F_{12} + \frac{1}{1/A_1F_{1R} + 1/A_2F_{2R}} \right] \sigma(T_{41} - T_{42})$$

written more simply as

$$Q_{12} = A_1 \overline{F}_{12} \sigma(T^4_{\;1} - T^4_{\;2})$$

With $A_1\overline{F}_{12} = A_1F_{12} + \left[\dfrac{1}{\dfrac{1}{A(1-F_{12})} + \dfrac{1}{A_2(1-F_{21})}}\right]$ (4.58)

Wherein F_{1R} and F_{2R} have been written as

$F_{1R}=1-F_{12}$

$F_{2R}=1-F_{21}$

Since the third surface completely surrounds the other two surfaces. Eqs. (4.58), on implication leads to

$$\overline{F}_{12} = \frac{A_2 - A_1(F_{12})^2}{A_1 - A_2 - 2A_1F_{12}}$$

(iii) Radiance Exchange between Infinite Parallel Gray Planes

In a system comprising two infinite parallel planes. A_1 and A_2 are equal: and the shape factor is 9unity since all the radiation leaving one plane reaches the other. The network is the same as in Fig.4.12 and the heat flow, Q_{12}, is obtained from Eqn. (4.57) as

$$Q_{12} = \frac{A\sigma(T^4_1 - T^4_2)}{\dfrac{1}{\epsilon_1} + \dfrac{1}{\epsilon_2} - 1}$$ (4.59)

(iv) Radiation Exchange between Large Concentric Cylinders or Spheres

In this case $F_{12} = 1.0$. The network is again the same as in Fig. 4.12 and the heat flow may be obtained from Eqn. (4.57) as

$$Q_{12} = \frac{A_1\sigma(T^4_1 - T^4_2)}{1/\epsilon_1 + (A_1/A_2)(1/\epsilon_2 - 1)}$$

(v) Radiation Exchange for Three Gray Surfaces

The network for three gray surfaces which see each other and nothing else is shown in Fig. The heat flow expressions are as follows.

$$Q_{12} = \frac{J_1 - J_2}{1/A_1F_{12}}$$

$$Q_{12} = \frac{J_1 - J_2}{1/A_1 F_{13}}$$

$$Q_{23} = \frac{J_1 - J_2}{1/A_2 F_{123}}$$

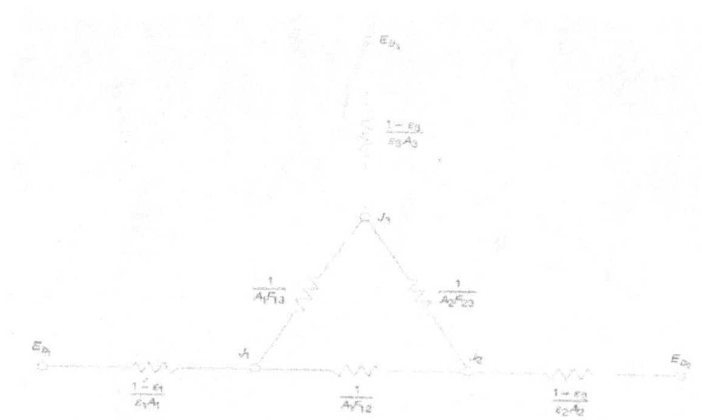

Fig. 4.15: Radiation Network for Three Gray Surfaces

The values of Q_{12} etc, are determined from the values of the radiosities which must be calculated first. The commonly used method is an application of Kirchoff's law of d.c. circuit theory. Which states that the sum of the currents entering a node is zero.

(vi) Radiation Exchange for Two Gray Surfaces Connected by a Reradiating Surface

The network for this system is shown in Fig. 4.16 (a). As explained for the case (ii) above, the system may be considered as a series parallel circuit as shown in Fig. 4.16 (b). The resistance of the circuit marked by dotted lines has been shown, as in Eqn. (4.58), to be

$$\frac{1}{A_1 F_{12}} = \left(\frac{A_1 + A_2 - 2A_1 F_{12}}{A_2 - A_1 (F_{12})^2} \right)$$

The total resistance of the system is therefore

$$R_f = \frac{1- \epsilon_1}{A_1 \epsilon_1} + \frac{1}{A_1} \left(\frac{A_1 + A_2 - 2A_1 F_{12}}{A_2 - A_1 (F_{12})^2} \right) + \frac{1- \epsilon_2}{A_2 \epsilon_2}$$

So that

$$Q_{12} = \frac{E_{b1} - E_{b2}}{R_T}$$

$$= \frac{A_1\sigma(T_1^4 - T_2^4)}{\left(\dfrac{1}{\epsilon_1} - 1\right) + \dfrac{A_1}{A_2}\left(\dfrac{1}{\epsilon_2} - 1\right) + \dfrac{A_1 + A_2 - 2A_1 F_{12}}{A_2 - A_1(F_{12})^2}}$$

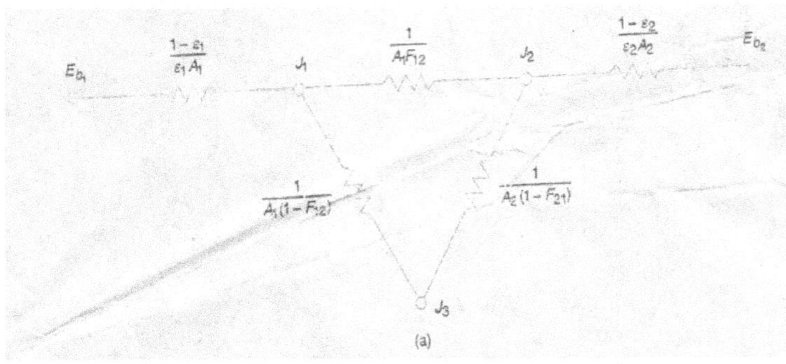

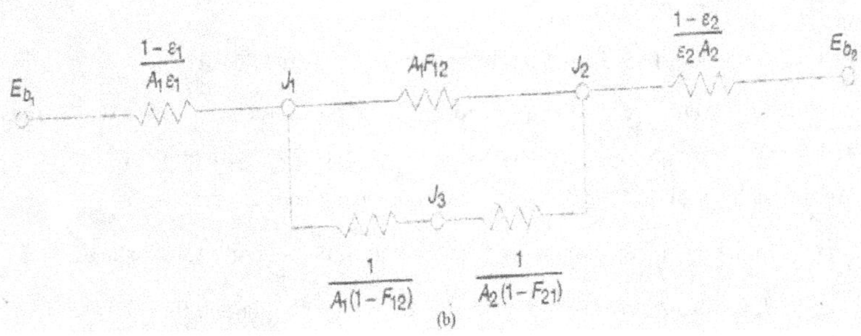

Fig. 4.16: Radiation Network for Two Gray Surfaces Connected by a Reradiating Surface

If $A_1 = A_2 = A$, then

$$\overline{F}_{12} = \frac{1 + F_{12}}{2}$$

And

$$Q_{12} = \frac{A\sigma(T_1^4 - T_2^4)}{\left(\dfrac{1}{\epsilon_1} + \dfrac{1}{\epsilon_2}\right) - 2 + \left(\dfrac{2}{1 + F_{12}}\right)}$$

(vii) Radiation Exchange for Four Gray Surfaces

The network for an enclosure consisting of four gray surfaces is shown in Fig. The heat flow from different surfaces can be estimated by considering each mode separately and applying Kirchhoff's law of circuit theory there.

The net heat flow from surface 1 is

$$Q = \frac{E_{b1} - J_1}{(1 - \epsilon_1)/A_1\,\epsilon_2} = \frac{J_1 - J_2}{1/A_1 F_{12}} + \frac{J_1 - J_3}{1/A_1 F_{13}} + \frac{J_1 - J_4}{1/A_1 F_{14}}$$

$$= A_1 F_{12}(J_1 - J_2) + A_1 F_{12}(J_1 - J_3) + A_1 F_{14}(J_1 - J_4)$$

Similar expressions can be written for other surfaces too.

The electrical network analogy serves as a very powerful tool for solving complex radiation exchange problems.

4.17. Formula Used in the Radiation Law Problems

S.NO	DESCRIPTION	FORMULA	HMT DB Pg.No
1	E missive power (E_b) Total emissive power Heat flux , (q) Irradiation,(G)	$E_b = \sigma T^4\ \dfrac{W}{m^2}$ σ - Stefan Boltzmann constant σ – 5.67* 10^{-8} W/m²k⁴ T – surface temperature –k	82
2	Wine's law	$\lambda_{max}T$ = 2898 µmk $\lambda_{max}T$= 2.9* 10^{-3} mk λ_{max} - maximum wave length (m)	82
3	Monochromatic emissive power (or) spectral emissive power (E_{bx})	$E_{bx} = \dfrac{C1\lambda^{-5}}{\rho^{\left[\frac{C_2}{\lambda T}\right]} - 1}$ C1 = 0.374* 10^{-15} W/m² C2 = 14.4* 10^{-3} mk X – wave length (µ) or 10^{-6} m	82
4	Maximum emissive power (E_{bx})max	$(E_{bx})max = C_4 T^5$ C4=1.307* 10^{-5}	82
5	Intensity of Radiation (I_n) = I$_\theta$	$I_n = \dfrac{E_b}{\pi}$	82
6	Total emissive power of a real surfaces (E_b) real	$(E_b)_{real} = E\sigma T^4$ E-Emissivity	82
7	Absorptivity (α) Reflectivity (e) Transmissivity (τ)	α =Radiation absorbed/Incident Radiation e =Radiation Reflected/Incident Radiation τ =Radiation Transmitted/Incident Radiation	82

4.18.   Problem based on the Radiation Property

1.   Assuming sun to be black body emitting radiation radiation with maximum intensity at $\lambda=0.5\mu$.

Calculate its surface temperature and emissive power.

Step 1

Wave length $\lambda=0.5 \mu = 0.5 \times 10^{-6}$m

$$[\therefore 1 \mu 10^{-6}\text{m}]$$

Step 2

To Find

1.   Surface temperature
2.   Emissive power

Step 3

1.   According to wien's displacement law

$\lambda_{max} . T = 2.9 \times 10^{-3}$mk

{From HMTDB P.No.82}

$0.5 \times 10^{-6} \times T = 2.9 \times 10^{-3}$

$T = 5800$k

Surface temperature $T = 5800$ k

Step 4

1.   According to Stefan – Boltzman law,

Emissive power $E_b = \sigma.T^4$

{From HMTDB P.No.82}

$E_b = 5.67 \times 10^{-8} \times (5800)^4$

$E_b = 64.1 \times 10^6 \dfrac{W}{m^2}$ $[\therefore \sigma = 5.6 \times 10^{-6} \dfrac{W}{m^2 K^4}]$

$E_b = 64.1 \ 10^6$ w/m²

Step 5

Surface temperature, $T = 5800$k

Emissive Power, $E_b = 64.1 \times 10^6$w/m²

2. It is observed that the intensity of the radiation emitted by the sun is maximum at a wave length of 0.5μ. Assuming the sun to be a black body. Estimate its surface temperature and emissive power.

Step 1

Given

According to wien's displacement law,

$\lambda_{max}^x \; T = 0.289 \times 10^{-2} mk$

$$T = \frac{0.289 \times 10^{-2}}{\lambda} = \frac{0.289 \times 10^{-2}}{0.5 \times 10^{-6}}$$

T = 5780 k

Step 2

According to the Stefan boltzman law,

$(E_b)_{sun} = \sigma.T^4$

$\qquad\qquad = 0.567 \times 10^{-7} \times (5780)^4$

$(E_b)_{sun} = 63.3 \dfrac{mw}{m^2}$

Step 3

Result

1. T = 5780 k

2. $(E_b)_{sun} = 63.3 \dfrac{mw}{m^2}$

3. An optical pyrometer is an instrument used to measure the high temperatures. It works on the principle of comparing the radiation leaving the body whose temperature is to be measured with that of an incondesent filament. The instrument is calibrated in such a way that it measures the temperatures which a black body would have if it were emitting radiation equal to that of the body under investigation. A pyrometer records the temperature of a body as 1400°C with a red light filter (λ=0.65μ). Find the true temperature of the body if its emissivity at 0. 65μ is 0.6.

Step 1

For a black body Planck's law,

$$E_b \lambda = \frac{2\pi C_1}{\lambda^5 \left[\exp\left(\dfrac{C_2}{\lambda T_b} \right) - 1 \right]}$$

Where T_b is the temperature of the body.

For a gray body,

$$E_\lambda = \frac{2\pi \, \varepsilon_\lambda \, C_1}{\lambda^5 \left[\exp\left(\dfrac{C_2}{\lambda_T} \right) - 1 \right]}$$

Step 2

Where, T is the absolute temperature of the gray body and at $E_{b\lambda} = E_\lambda$. Tb will be the temperature indicated by the pyrometer.

Here, $\dfrac{C_2}{\lambda T_0} = \dfrac{1.437 \times 10^{-2}}{0.65 \times 10^{-6} \times 1673} = 13.23 > 1$

Step 3

The denominator of the above formula can be taken to be, $\lambda^5 \exp\,(C^2/\lambda T)$

So, equating $E_b\lambda$ with E_λ

$$\frac{1}{T} = \frac{1}{T_b} - \frac{\lambda}{C_2} \ln \frac{1}{\varepsilon\lambda}$$

$$T = \frac{1}{\dfrac{1}{T_b} - \dfrac{\lambda}{C_2} l n \dfrac{1}{\varepsilon_\lambda}}$$

$$= \frac{1}{\dfrac{1}{1673} - \dfrac{0.65 \times 10^{-4}}{1.439 \times 10^{-2}} l n \left(\dfrac{1}{0.6} \right)} = 1740\,k$$

Step 4

Temperature of the body = 1740 – 273 = 1467°C

Description	Formula	
Square plate	$$Q_{12} = \dfrac{\sigma\left[T_1^4 - T_2^4\right]}{\dfrac{1-\varepsilon_1}{A_1\varepsilon_1} + \dfrac{1}{A_1F_{12}} + \dfrac{1-\varepsilon_2}{A_2\varepsilon_2}}$$ For black body $\varepsilon_1 = \varepsilon_2 = 1$ $Q_{12} = \sigma\,[T_1^4 - T_2^4]\ \times A_1 F_{12}$ Where F_{12} – shape factor for square plate x-axis = smaller side / distance between plates Curve → 2 [for square level)	Find F12 from HMT
Circular disc	$$Q_{12} = \dfrac{\sigma\left[T_1^4 - T_2^4\right]}{\dfrac{1-\varepsilon_1}{A_1\varepsilon_1} + \dfrac{1}{A_1F_{12}} + \dfrac{1-\varepsilon_2}{A_2\varepsilon_2}}$$ where	
	F_{12} – shape factor for disc $X\text{-axis} - \dfrac{\text{Diameter}}{\text{Dis}\tan\text{ce between discs}}$ Curve → 1 [for disc]	
	$$Q_{12} = \dfrac{\sigma\left[T_1^4 - T_2^4\right]}{\dfrac{1-\varepsilon_1}{A_1\varepsilon_1} + \dfrac{1}{A_1F_{12}} + \dfrac{1-\varepsilon_2}{A_2\varepsilon_2}}$$ $X\text{ axis} \to \dfrac{\text{Longerside}}{\text{Dis}\tan\text{ce}}\ \dfrac{L}{D}$ $\text{Curve} \to \dfrac{\text{Breath}}{\text{Dis}\tan\text{ce}}\ \dfrac{B}{D}$	

Step 5

Result

Temperature of the body = 1467ºC

4.19. Problems based on the Heat Exchange between the Two Surface

1. Two black square plates of size 2 x 2m are placed parallel to each other at a distance of 0.5m one plate is maintained at a temperature of 1000°C and the other at 500°C. Find the heat exchange between the plates.

Step 1

Given

Area, A = 2x2 =4m²

T_1 = 1000°C + 273 = 1273K

T_2 = 500°C + 273 = 773 K

Distance = 0.5m.

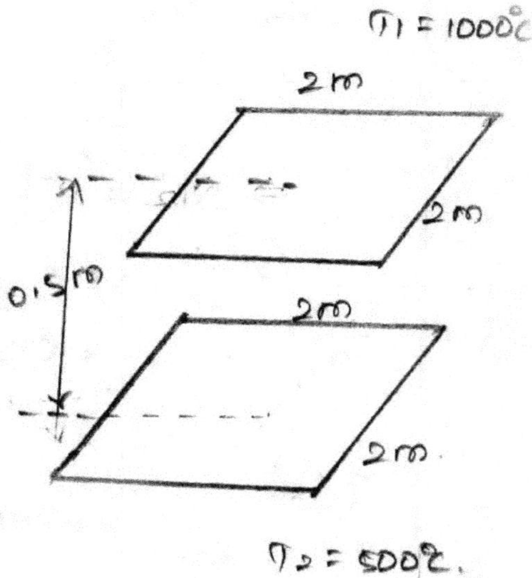

Step 2

Heat transfer (Q)

Step 3

Heat transfer by radiation general equation is,

$$Q_{12} = \frac{\sigma\left[T_1^4 - T_2^4\right]}{\dfrac{1-\varepsilon_1}{A_1\varepsilon_1} + \dfrac{1}{A_1 F_{12}} + \dfrac{1-\varepsilon_2}{A_2\varepsilon_2}}$$

For black body,

$\varepsilon_1 = \varepsilon_2 = 1$

$\Rightarrow Q_{12} = \sigma\left[T_1^4 - T_2^4\right] \times A_1 F_{12}$

$= 5.67 \times 10^{-8}\left[(1273)^4 - (7773)^4\right] \times 4 \times F_{12}$

$Q_{12} = 5.14 \times 10^5\, F_{12}$ $\qquad\qquad \rightarrow (1)$

Where, F_{12} – Shape factor for square plates

Step 4

$$x_{axis} = \frac{\text{Smaller Side}}{\text{Dis}\tan\text{ce between plates}} \qquad \text{\{HMT DB P.no.90\}}$$

$= {}^{2}\!/_{0.5} \quad = 4$

$x_{axis} = 4$

Step 5

Curve $\rightarrow 2$ $\qquad\qquad$ {Given is square plates}

x_{axis} value is 4, curve is 2, so, corresponding x axis value is 0.62.

$F_{12} = 0.62$

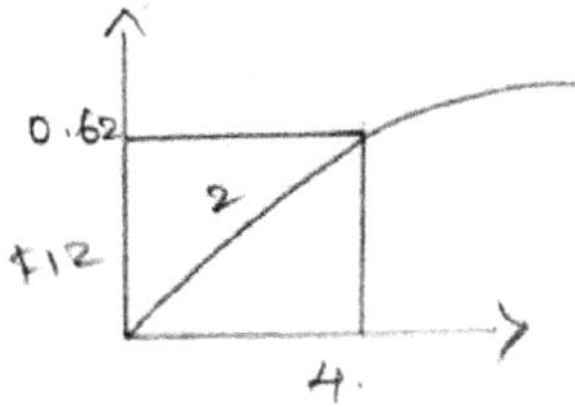

(1) $\Rightarrow$ $\qquad Q_{12} = 5.14 \times 10^5 \times 0.62$

$Q_{12} = 3.18 \times 10^5\, \text{w}$

Step 6

Heat transfer $Q_{12} = 3.18 \times 10^5\, \text{W}$

2. Two circular discs of diameter 0.3m each are placed parallel to each other at a distance of 0.2m. one disc is maintained at a temperature of 750°C and the other at 350°C and their corresponding emissivity are 0.3 and 0.6. Calculate the heat exchange between the discs.

Step 1

Given

$D_1 = 0.3$m

$D_2 = 0.3$m

$A_1 = A_2 = \dfrac{\pi}{4} D^2 = \dfrac{\pi}{4}(0.3)^2$

$A_1 = A_2 = 0.070$ m²

$T_1 = 750$°C + 273 = 1023k

$T_2 = 350$°C + 273 = 623k

$\varepsilon_1 = 0.3$

$\varepsilon_2 = 0.6$

Distance between discs = 0.2m

Step 2

Heat exchange between discs (Q)

Step 3

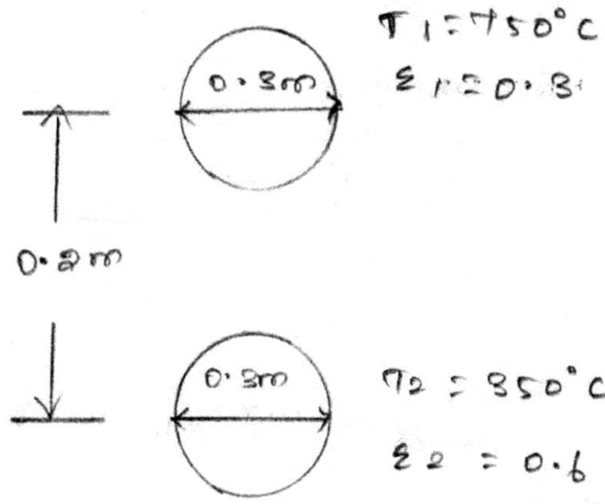

Step 4

Heat transfer by radiation general equation is,

$$Q_{12} = \frac{\sigma\left[T_1^4 - T_2^4\right]}{\dfrac{1-\varepsilon_1}{A_1\varepsilon_1} + \dfrac{1}{A_1 F_{12}} + \dfrac{1-\varepsilon_2}{A_2\varepsilon_2}} = \frac{5.67 \times 10^{-8}\left[(1023)^4 - (623)^4\right]}{\dfrac{1-0.3}{0.070 \times 0.3} + \dfrac{1}{0.070\, F_{12}} + \dfrac{1-0.6}{0.070 \times 0.6}}$$

$$Q_{12} = \frac{5.35 \times 10^4}{42.85 + \dfrac{1}{0.070\, F_{12}}}$$

Where F_{12} = shape factor for disc for refer HMT DP 90.

$$x_{axis} = \frac{Diameter}{Dis\tan ce\ between\ discs}$$

$$x_{axis} = \frac{0.3}{0.2} = 1.5$$

Step 5

Curve $\rightarrow$ 1 (Since given is disc)

x_{axis} value is 1.5 and curve is 1.

So, corresponding x_{axis} value is 0.28

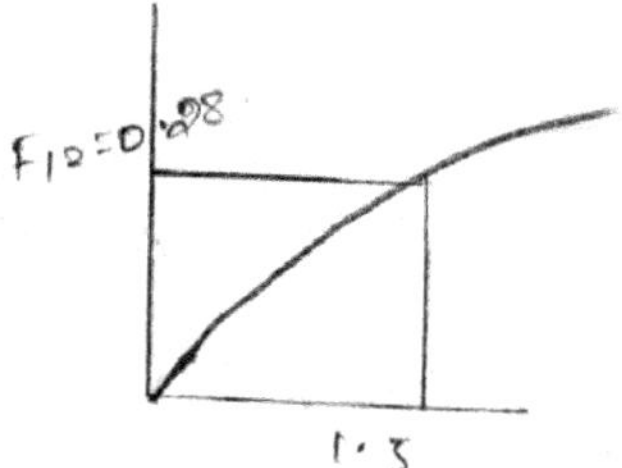

$F_{12} = 0.28$

$$(1) \Rightarrow Q_{12} = \frac{5.35 \times 10^4}{42.85 + \dfrac{1}{0.070 \times 0.28}}$$

Q_{12} = 569.9 w

Step 6

Result

Heat transfer Q = 569.9w

3. Two parallel rectangular surfaces 1m x 2m are opposite to each other at a distance of 4m. The surfaces are black and at 300ºC and 200ºC. Calculate the heat exchange by radiation between the two surfaces.

Step 1

Given

Area, A = 1 x 2 = 2m²

Distance = 4m

T_1 = 300ºC + 273 = 573 k

T_2 = 200ºC + 273 = 473 k

Step 2

Heat exchange (Q_{12})

Step 3

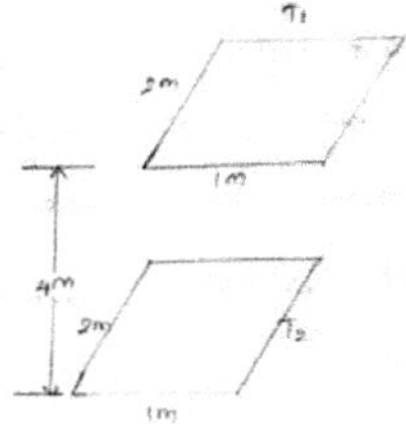

Step 4

$$Q_{12} = \frac{\sigma\left[T_1^4 - T_2^4\right]}{\dfrac{1-\varepsilon_1}{A_1\varepsilon_1} + \dfrac{1}{A_1F_{12}} + \dfrac{1-\varepsilon_2}{A_2\varepsilon_2}}$$

For Black Surface,

$\varepsilon_1 = \varepsilon_2 = 1$

$\Rightarrow Q_{12} = \sigma \left[T_1^4 - T_2^4\right] \times A_1F_{12}$

F_{12} = Shape factor for parallel rectangular

$$X = \frac{L}{D} = \frac{\text{Longer side}}{\text{Dis}\tan\text{ce}} = \frac{2}{4} = 0.5$$

$$Y = \frac{B}{D} = \frac{1}{4} = 0.25$$

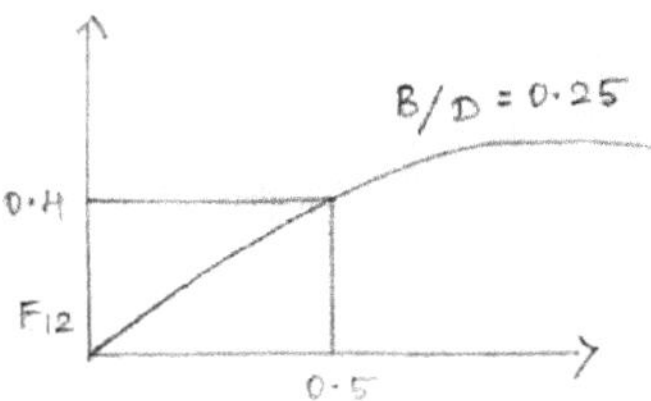

From graph, we know that,

$F_{12} = 0.04$

$(1) \Rightarrow Q_{12} = 5.67 \times 10^{-8} \times [(573)^4 - (473)^4] \times 2 \times 0.04$

$Q_{12} = 261.9w$

Step 5

Heat exchange, $Q_{12} = 261.9$ w

4.20. Formula Used in the Radiation Shield

Radiation Shield

Description	Formula	Where
Heat exchange for plate Two large parallel plate T_1 T_2 ε_1 ε_2 Two large parallel plate with radiation shield 1 3 2 plate 1 plate 2 ε_1 ε_2 T_1 T_2 T_3 Radiation shield, ε_8	$Q_{12} = \bar{\varepsilon}\,\sigma\,A\left[T_1^4 - T_2^4\right]$ $\bar{\varepsilon} = \dfrac{1}{\dfrac{1}{\varepsilon_1} + \dfrac{1}{\varepsilon_2} - 1}$ $Q_{12} = \bar{\varepsilon}\,\sigma\,A\left[T_1^4 - T_2^4\right]$ $\bar{\varepsilon} = \dfrac{1}{\dfrac{1}{\varepsilon_1} + \dfrac{1}{\varepsilon_2} - 1}$ $Q_{13} = \bar{\varepsilon}\,\sigma\,A\left[T_1^4 - T_3^4\right]$ $\bar{\varepsilon} = \dfrac{1}{\dfrac{1}{\varepsilon_1} + \dfrac{1}{\varepsilon_2} - 1}$ $Q_{32} = \bar{\varepsilon}\,\sigma\,A\left[T_3^4 - T_2^4\right]$ $\bar{\varepsilon} = \dfrac{1}{\dfrac{1}{\varepsilon_1} + \dfrac{1}{\varepsilon_2} - 1}$	σ = Stofapn – Boltzman = $5.67 \times 10^{-8} w/m^2$ ε_1=Emissivity of ε_1=Emissivity of T_1 = Temperature of T_2 = Temperature of Q = Heat exchange in

Description	Formula	Where
For cylinder (or) sphere Two large concentric cylinder (or) sphere with insulation	$Q_{12} = \bar{\varepsilon}\,\sigma\,A\left[T_1^4 - T_2^4\right]$ $\bar{\varepsilon} = \dfrac{1}{\dfrac{1}{\varepsilon_1} + \dfrac{1}{\varepsilon_3} - 1}$	For cylinder Area $A = 2\pi r$ For sphere Area $A = 4\pi r^2$
Without insulation	$Q_1 = \bar{\varepsilon}\,\sigma\,A\left[T_1^4 - T_2^4\right]$	
	$Q_{11} = \dfrac{A\sigma\,[T_1^4 - T_2^4]}{\dfrac{1}{\varepsilon_1} + \dfrac{1}{\varepsilon_3} + \dfrac{2n}{\varepsilon_s} - (n+1)}$ $Q_{12} = \dfrac{A\sigma\,[T_1^4 - T_2^4]}{\dfrac{1}{\varepsilon_1} + \dfrac{1}{\varepsilon_2} - 1}$ Find number of shield from $\dfrac{Q_{12}}{Q_{1n}}$ ratio	N = number ε_s = Emissivity shield
Two parallel plate with two radiation shield (i) Without raidation	$Q_{12} = \bar{\varepsilon}\,\sigma\,A\left[T_1^4 - T_2^4\right]$ $\bar{\varepsilon} = \dfrac{1}{\dfrac{1}{\varepsilon_1} + \dfrac{1}{\varepsilon_3} - 1}$	
With ratiation shields	$Q_{1,3a} = \bar{\varepsilon}\,\sigma\,A\left[T_1^4 - T_2^4\right]$ $\bar{\varepsilon} = \dfrac{1}{\dfrac{1}{\varepsilon_1} + \dfrac{1}{\varepsilon_{3a}} - 1}$ $Q_{3b,2} = \bar{\varepsilon}\,\sigma\,A\left[T_3^4 - T_2^4\right]$ $\bar{\varepsilon} = \dfrac{1}{\dfrac{1}{\varepsilon_1} + \dfrac{1}{\varepsilon_{3a}} - 1}$ Find T_3 value from equality $Q_{1,3a} = Q_{3b2}$	

4.21. Problems based on the Radiation Shield

1. Calculate the net radiant heat exchange per m² area for two large parallel plates at temperature of 427°C and 27°C respectively. $\varepsilon_{(hot\ plate)}$ = 0.9, $\varepsilon_{(cold\ plate)}$ = 0.6. If a polished aluminium shield is placed between them, find the percentage of reduction in the heat transfer. $\varepsilon_{(Shield)}$ = 0.4.

Step 1

Given

T_1 = 427°C + 273 = 700k

T_2 = 27°C + 273 = 300k

ε_1 = 0.9

ε_2 = 0.6

ε_3 = 0.4

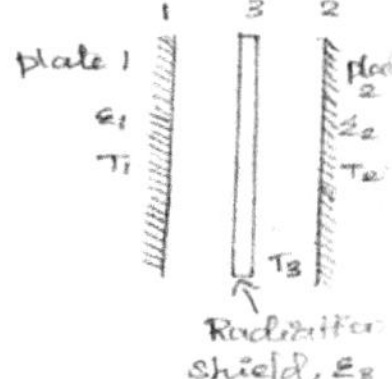

Step 2

1. Net radiant heat exchange per m₂ area
2. Percentage of reduction in the heat transfer.

Step 3

Case (i) Heat Transfer without Radiation Shield

$$Q_{12} = \bar{\varepsilon}\,\sigma\, A\left[T_1^4 - T_2^4 \right]$$

$$\bar{\varepsilon} = \cfrac{1}{\dfrac{1}{\varepsilon_1} + \dfrac{1}{\varepsilon_2} - 1}$$

$$\bar{\varepsilon} = 0.5625$$

$$\Rightarrow Q_{12} = 0.5625 \times 5.67 \times 10^{-8} \times A \times [700^4 - 300^4]$$

$$\frac{Q_{12}}{A} = 7.39 \times 10^3 \text{ W/m}^2$$

Step 4

Heat Transfer with Radiation Shield

Heat exchange between plate 1 and radiation shield 3 given by,

$$Q_{12} = \bar{\varepsilon}\, \sigma\, A\left[T_1^4 - T_3^4\right]$$

$$\bar{\varepsilon} = \cfrac{1}{\dfrac{1}{\varepsilon_1} + \dfrac{1}{\varepsilon_3} - 1}$$

$$Q_{13} = \cfrac{\sigma\, A\left[T_1^4 - T_3^4\right]}{\dfrac{1}{\varepsilon_1} + \dfrac{1}{\varepsilon_3} - 1}$$

Where, $\bar{\varepsilon} = \cfrac{1}{\dfrac{1}{\varepsilon_1} + \dfrac{1}{\varepsilon_3} - 1}$

$$Q_{13} = \cfrac{\sigma\, A\left[T_3^4 - T_2^4\right]}{\dfrac{1}{\varepsilon_3} + \dfrac{1}{\varepsilon_2} - 1}$$

$\therefore Q_{13} = Q_{32}$

Step 5

$$\cfrac{\sigma A\left[T_1^4 - T_3^4\right]}{\dfrac{1}{\varepsilon_1} + \dfrac{1}{\varepsilon_3} - 1} = \cfrac{\sigma A\left[T_3^4 - T_2^4\right]}{\dfrac{1}{\varepsilon_3} + \dfrac{1}{\varepsilon_2} - 1}$$

$$\cfrac{\left(T_1^4 - T_3^4\right)}{\dfrac{1}{\varepsilon_1} + \dfrac{1}{\varepsilon_3} - 1} = \cfrac{\left(T_3^4 - T_2^4\right)}{\dfrac{1}{\varepsilon_3} + \dfrac{1}{\varepsilon_2} - 1}$$

$$\cfrac{(700)^4 - T_3^4}{\dfrac{1}{0.9} + \dfrac{1}{0.4} - 1} = \cfrac{T_3^4 - (300)^4}{\dfrac{1}{0.4} + \dfrac{1}{0.6} - 1}$$

$7.8 \times 10^{11} = 5.77\, T_3^4$

$T_3^4 = 1.353 \times 10^{11}$

Radiation shield temperature,

$T_3 = 606.55\text{k}$

Heat transfer

With radiation $Q_{13} = \cfrac{\sigma\, A\left[T_1^4 - T_3^4\right]}{\dfrac{1}{\varepsilon_1} + \dfrac{1}{\varepsilon_3} - 1}$

$$\text{Shield} = \frac{5.67 \times 10^{-8} \times A \times \left[(700)^4 - (606.55)^4 \right]}{\dfrac{1}{0.9} + \dfrac{1}{0.4} - 1}$$

$$\frac{Q_{13}}{A} = 2.27 \times 10^3 \ \text{W}\!\big/\!\text{m}^2$$

Reduction in heat loss

Due to radiation shield $= \dfrac{Q_{without} - Q_{with}}{Q_{without\ shield}}$

$$= \frac{Q_{12} - Q_{13}}{Q_{12}}$$

$$= \frac{7.39 \times 10^3 - 2.27 \times 10^3}{7.39 \times 10^3} = 0.692 = 69.2\%$$

Step 6

1. Net radiant heat exchange (Without shield) $\dfrac{Q_{12}}{A} = 7.39 \times 10^3 \ \text{W}\!\big/\!\text{m}^2$

2. Percentage of reduction in the heat transfer due to shield = 69.2%

2. Emissivities of the two large parallel planes maintained at 800°C and 300°C are 0.3 and 0.5 respectively. Find the net radiant heat transfer per square meter for these plates.

Step 1

Given

$T_1 = 800°C + 273 = 1073k$

$T_2 = 300°C + 273 = 573k$

$\varepsilon_1 = 0.3$

$\varepsilon_2 = 0.5$

Step 2

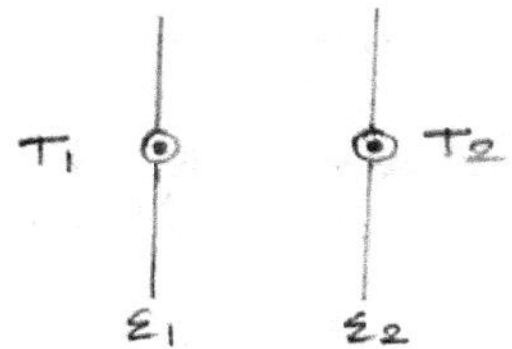

Step 3

To Find

Heat Exchange between two large parallel plate is given by,

$$Q = \bar{\varepsilon}\sigma A\left[T_1^4 - T_2^4\right] = \cfrac{1}{\dfrac{1}{\varepsilon_1} + \dfrac{1}{\varepsilon_2} - 1} = \cfrac{1}{\dfrac{1}{0.3} + \dfrac{1}{0.5} - 1}$$

$$\bar{\varepsilon} = 0.23$$

$$Q = 0.23 \times \sigma \times A \times (T_1^4 - T_2^4)$$

$$\frac{Q}{A} = 0.23 \times 5.67 \times 10^{-8} \times \left[1073^4 - 573^4\right]$$

$$\frac{Q}{A} = 15.8 \times 10^3 \; \text{W}/\text{m}^2$$

Step 4

$$\frac{Q}{A} = 15.8 \times 10^3 \; \text{W}/\text{m}^2$$

3. Emissivities of two large parallel plates maintained at T_1 k and T_2 k are 0.6 and 0.6 respectively. Heat transfer is reduced 75 times when a polished aluminium radiation shields of emissivity 0.04 are placed in between them. Calculate the number of shields required.

Step 1

Given

$\varepsilon_1 = 0.6$

$\varepsilon_2 = 0.6$

Heat transfer is reduced = 75 times

Emissivity of radiation shield = $\varepsilon_s = \varepsilon_3 = 0.04$

Step 2

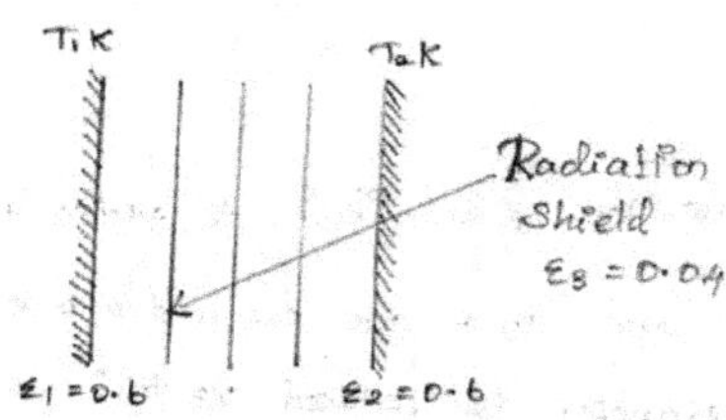

Step 3

Number of shields required.

Step 4

Heat transfer with n shield is given by,

$$Q_{1n} = \frac{A\sigma\left[T_1^4 - T_2^4\right]}{\frac{1}{\varepsilon_1} + \frac{1}{\varepsilon_2} + \left(\frac{2n}{\varepsilon s}\right) - (n+1)} \qquad \rightarrow (1)$$

Heat transfer without shield i.e., n = 0

$$(1) \Rightarrow Q_{12} = \frac{A\sigma\left[T_1^4 - T_2^4\right]}{\frac{1}{\varepsilon_1} + \frac{1}{\varepsilon_2} - 1} \qquad \rightarrow (2)$$

Heat transferred is reduced 75 times,

$$\frac{Q_{without\,shield}}{Q_{with\,shield}} = 75$$

$$\frac{Q_{12}}{Q_{1n}} = 75$$

$$\frac{(2)}{(1)} \Rightarrow \frac{A\sigma\left[T_1^4 - T_2^4\right]}{\frac{1}{\varepsilon_1} + \frac{1}{\varepsilon_2} - 1} = 75$$

$$\frac{A\sigma\left[T_1^4 - T_2^4\right]}{\frac{1}{\varepsilon_1} + \frac{1}{\varepsilon_2} + \frac{2n}{\varepsilon_s} - (n+1)}$$

$$\frac{\frac{1}{\varepsilon_1} + \frac{1}{\varepsilon_2} + \frac{2n}{\varepsilon s} - (n+1)}{\frac{1}{\varepsilon_1} + \frac{1}{\varepsilon_2} - 1} = 75$$

$$\frac{\frac{1}{0.6} + \frac{1}{0.6} + \frac{2n}{0.04} - (n+1)}{\frac{1}{0.6} + \frac{1}{0.6} - 1} = 75$$

$$\frac{3.33 + 50n - n - 1}{2.33} = 75$$

$$50n - n - 1 = 171.67$$

49n – 1 = 171.67

49n = 172.67

n = 3.52 = 4

n = 4

Step 5

Number of shields required n = 4 nos

4. Two large parallel planes at 800k and 600k have emissivities of 0.5 and 0.8 respectively. A radiation shield having an emissivity of 0.1 on one side and on emissivity of 0.05 on the other side is placed between the plates calculate the heat transfer rate by radiation per square meter with and without radiation shield. Comment on the results.

Step 1

Given

T_1 = 800k

T_2 = 600k

ε_1 = 0.5

ε_2 = 0.8

ε_{3a} = 0.1

ε_{3b} = 0.05

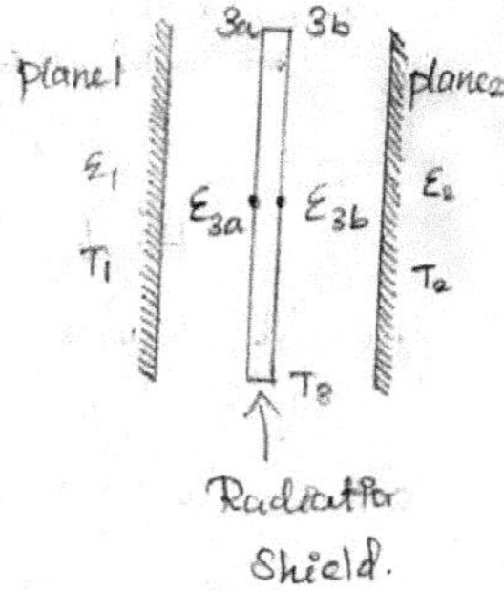

To Find

1. Heat transfer rate per square meter without radiation shield

2. Heat transfer rate per square metre with radiation shield

3. Comment on the results.

Step 3

Solution

Case 1

Heat transfer without radiation shield

Heat exchange between two parallel plates without radiation shield is given by

$$Q_{12} = \bar{\varepsilon}\, \sigma\, A\left[T_1^4 - T_2^4\right]$$

(from eqn.no. (4.28))

Where, $\bar{\varepsilon} = \dfrac{1}{\dfrac{1}{\varepsilon_1} + \dfrac{1}{\varepsilon_2} - 1} = \dfrac{1}{\dfrac{1}{0.5} + \dfrac{1}{0.8} - 1}$

$\bar{\varepsilon} = 0.444$

$Q_{12} = 0.444 \times 5.67 \times 10^{-8} \times A \times [(800)^4 - (600)^4]$

$$\dfrac{Q_{12}}{A} = 7.048 \times 10^3 \; {}^{\text{w}}\!\big/_{\text{m}^2}$$

Heat transfer

Without radiation $\qquad \dfrac{Q_{12}}{A} \qquad = 7.048 \times 103 \text{ w/m}$

Shield

Step 4

Case 2

Heat transfer between plate 1 and radiation shield 3a is given by

$$Q_{1,3a} = \bar{\varepsilon}\, \sigma\, A\left[T_1^4 - T_3^4\right]$$

To here,

$$\bar{\varepsilon} = \dfrac{1}{\dfrac{1}{\varepsilon_1} + \dfrac{1}{\varepsilon_{3a}} - 1}$$

$$Q_{1,3a} = \dfrac{\sigma\, A\left[T_1^4 - T_3^4\right]}{\dfrac{1}{\varepsilon_1} + \dfrac{1}{\varepsilon_{3a}} - 1} \qquad\qquad \rightarrow (2)$$

Heat exchange between radiation shield 3b and plate 2 is given by

$$Q_{3b,\,2} = \overline{\varepsilon}\,\sigma\,A\left[T_3^4 - T_2^4\right]$$

Where,

$$\overline{\varepsilon} = \cfrac{1}{\cfrac{1}{\varepsilon_{3b}} + \cfrac{1}{\varepsilon_2} - 1}$$

We know that,

$$Q_{1,3Q} = Q_{3b,\,2}$$

$$\Rightarrow \quad \cfrac{\sigma\,A\left[T_1^4 - T_3^4\right]}{\cfrac{1}{\varepsilon_1} + \cfrac{1}{\varepsilon_{3a}} - 1} = \cfrac{\sigma\,A\left[T_3^4 - T_2^4\right]}{\cfrac{1}{\varepsilon_{3b}} + \cfrac{1}{\varepsilon_2} - 1}$$

$$\Rightarrow \quad \cfrac{T_3^4 - T_2^4}{\cfrac{1}{\varepsilon_1} + \cfrac{1}{\varepsilon_{3a}} - 1} = \cfrac{T_3^4 - T_2^4}{\cfrac{1}{\varepsilon_{3b}} + \cfrac{1}{\varepsilon_2} - 1}$$

$$\Rightarrow \quad \cfrac{(800)^4 - T_3^4}{\cfrac{1}{0.5} + \cfrac{1}{0.1} - 1} = \cfrac{T_3^4 - (600)^4}{\cfrac{1}{0.05} + \cfrac{1}{0.8} - 1}$$

$$\Rightarrow \quad \cfrac{(800)^4 - T_3^4}{\cfrac{1}{0.5} + \cfrac{1}{0.1} - 1} = \cfrac{T_3^4 - (600)^4}{20.25}$$

$\Rightarrow \quad 20.25\,[(800)^4 - T_3^4] = 11\,[T_3^4 - (600)^4]$

$\Rightarrow \quad 8.29 \times 10^{12} - 20.25\,T_3^4 = 11 T_3^4 - 1.42 \times 10^{12}$

$\Rightarrow \quad 9.71 \times 10^{12} = 31.25\,T_3^4$

$\Rightarrow \quad T_3^4 = 3.1072 \times 10^{11}$

Radiation shield temperature,

$T_3 = 746.60k$

Step 5

Substitute T_3 value in equation (2) or (3)

Heat transfer

With radiation $Q_{1,3a} = \cfrac{\sigma\,A\left[T_1^4 - T_3^4\right]}{\cfrac{1}{\varepsilon_1} + \cfrac{1}{\varepsilon_{3a}} - 1}$

$$\text{Shield} = \frac{5.67 \times 10^{-8} \times A\left[(800)^4 - (746.60)^4\right]}{\dfrac{1}{0.5} + \dfrac{1}{0.1} - 1}$$

$$\frac{Q_{1,3a}}{A} = 509.74 \text{ w/m}^2$$

Heat transfer

With radiation $\dfrac{Q_{1,3a}}{A} = 509.74 \; {}^{W}\!\big/\!{}_{m^2}$

Shield

Reduction in heat

Transfer due to $= \dfrac{Q_{\text{without shield}} - Q_{\text{with shield}}}{Q_{\text{without shield}}}$

Radiation shield

$$= \frac{Q_{12} - Q_{1,3a}}{Q_{12}}$$

$$= \frac{7.048 \times 10^3 - 509.74}{Q_{12}}$$

$$= \frac{7.048 \times 10^3 - 509.74}{7.048 \times 10^3}$$

$$= 0.927$$

$$= 92.7\%$$

Step 6

Result

1. Heat transfer

    Without radiation $\quad \dfrac{Q_{12}}{A} = 7.048 \times 10^3 \; {}^{W}\!\big/\!{}_{m^2}$

    Shield

2. Heat transfer with

    Radiation shield $\quad \dfrac{Q_{1,3a}}{A} = 509.74 \; {}^{W}\!\big/\!{}_{m^2}$

4.22. Problem based on the Heat Loss between the Cylinder

1. A pipe of diameter 30cm, carrying steam runs in a large room and is exposed to air at a temperature of 25°C. The surface temperature of the pipe is 300°C. Calculate the loss of heat to surrounding per meter length of pipe due to thermal radiation. The emissivity of the pipe surface is 0.8.

    What would be the loss of heat due to radiation of the pipe which is enclosed in a 55 cm diameter brick of emissivity 0.91?

Step 1

Case 1

Given

Diameter of pipe, D_1 = 30cm = 0.30m

Surface temperature, T_1 = 300°C + 273 = 573k

Air temperature, T_2 = 25°C + 273 = 298k

Emissivity of the pipe, ε_1 = 0.8

Step 2

Case 2

Cuter diameter, D_2 = 55cm = 0.55m

Emissivity, ε_2 = 0.91

To Find

1. Loss of heat pea metre length (Q/L)
2. Reduction in heat loss.

Step 3

Solution

Case 1

Heat transfer, $Q = \varepsilon_1 \sigma A [T_1^4 - T_2^4] = \varepsilon_1 \times \sigma \times \pi D [T_1^4 - T_2^4]$

$[\because A = \pi DL]$

$Q = 0.8 \times 5.67 \times 10^{-8} \times \pi \times 0.30 \times L \times [(573)^4 - (298)^4]$

$Q/L = 4271.3 \ w/m$

Heat loss per metre length = 4271.3 w/m

Step 4

Case 2

When the 30cm diameter pipe is enclosed in a 55cm diameter pipe, heat exchange between two large concentric cylinder is given by

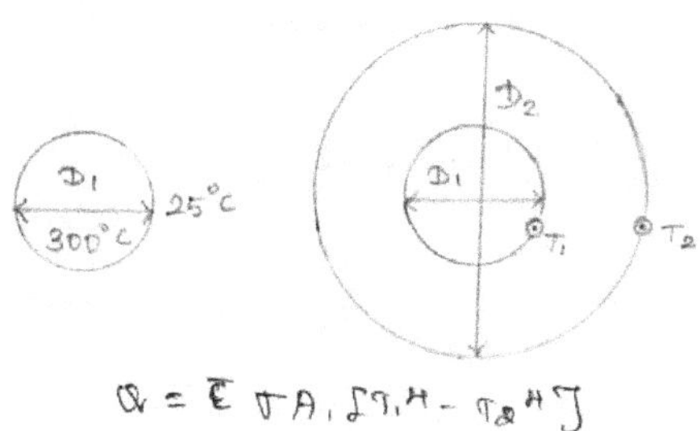

$$Q = \overline{\varepsilon} \sigma A1 \left[T_1^4 - T_2^4 \right]$$

Step 5

Where,

$$\overline{\varepsilon} = \cfrac{1}{\cfrac{1}{\varepsilon_1} + \cfrac{A1}{A2}\left[\cfrac{1}{\varepsilon_2} - 1\right]}$$

$$= \cfrac{1}{\cfrac{1}{0.8} + \cfrac{\pi D_1 L}{\pi D_2 L}\left(\cfrac{1}{0.91} - 1\right)}$$

$$= \cfrac{1}{\cfrac{1}{0.8} + \cfrac{D_1}{D_2}\left(\cfrac{1}{0.91} - 1\right)}$$

$$= \cfrac{1}{\cfrac{1}{0.8} + \cfrac{0.30}{0.55}\left(\cfrac{1}{0.91} - 1\right)}$$

$$\overline{\varepsilon} = 0.76$$

Substituting emissivity value in eqn (1)

$$(1) \Rightarrow Q \quad = 0.76 \times 5.67 \times 10^{-8} \times \pi \times D \times L_1 \times [(573)^4 - (298)^4]$$

$$= 0.76 \times 5.67 \times 10^{-8} \times \pi \times 0.30 \times [(573)^4 - (298)^4]$$

$$= 4057.8 \text{ w/m}$$

Step 6

Reduction in heat loss

$= 4271.3 - 4057.8$

$= 213.4$

Step 7

Result

1. Heat loss per metre length = 4271.3 w/m
2. Reduction in heat loss = 213.4 w/m

2. Calculate the heat lost by radiation per metre length of 8cm diameter pipe at 400°C and emissivity of 0.7, when

a) It is located in a large room with a red brick walls maintained at a temperature of 35°C.

b) It is enclosed in a 20cm diameter of red brick pipe maintained at a temperature of 50°C and emissivity of 0.9. Also find reduction in heat loss.

Step 1

Case 1

Given

Length, $L = 150$

Diameter of Pipe, $D_1 = 8cm = 0.08m$

Temperature, $T_1 = 400°C + 273$

$T_1 = 673k$

Emissivity, $\varepsilon_1 = 0.7$

Temperature, $T_2 = 35°C + 273$

$T_2 = 308k$

Step 2

Case 2

Diameter, $D_2 = 20cm$

$= 0.20m$

Temperature, $T_2 = 50°C + 273$

$= 323k$

Emissivity $\varepsilon_2 = 0.9$

Step 3

To Find

1. Heat lost by radiation
2. Reduction in heat loss

Solution

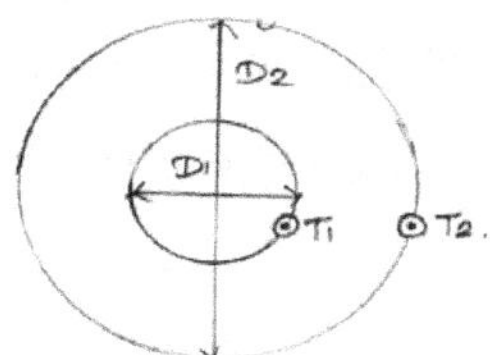

Step 4

Case 1

Heat Exchange $Q_1 = \varepsilon_1 \sigma A [T_1^4 - T_2^4]$

$\qquad = 0.7 \times 5.67 \times 10^{-8} \times \pi \times D_1 \times L \times [(673)^4 - (308)^4]$

$\qquad = 0.7 \times 5.67 \times 10^{-8} \times \pi \times 0.08 \times 1 \times [(673)^4 - (308)^4]$ (1)

$\qquad Q_1 = 1956.5$ w

Step 5

Case 2

Heat exchange between two large concentric cylinders is given by

$$Q_2 = \bar{\varepsilon} \sigma A1 \left[T_1^4 - T_2^4 \right] \qquad (2)$$

Where,

$$\bar{\varepsilon} = \cfrac{1}{\cfrac{1}{\varepsilon_1} + \cfrac{A1}{A2}\left[\cfrac{1}{\varepsilon_2} - 1\right]}$$

$$= \cfrac{1}{\cfrac{1}{\varepsilon_1} + \cfrac{\pi D_1 L}{\pi D_2 L}\left(\cfrac{1}{\varepsilon_2} - 1\right)}$$

$$= \cfrac{1}{\cfrac{1}{\varepsilon_1} + \cfrac{D_1}{D_2}\left(\cfrac{1}{\varepsilon_2} - 1\right)}$$

$$= \cfrac{1}{\cfrac{1}{0.7} + \cfrac{0.008}{0.2}\left(\cfrac{1}{0.9} - 1\right)}$$

$$\bar{\varepsilon} = 0.67\,w$$

$(2) \Rightarrow \quad Q \quad = 0.67 \times 5.67 \times 10^{-8} \times \pi \times D_1 \times L \times [(673)^4 - (323)^4]$

$\qquad\qquad = 0.67 \times 5.67 \times 10^{-8} \times \pi \times 0.08 \times 1 \times [(673)^4 - (323)^4]$

$\qquad Q_2 = 1854.7\,w$

Reduction in heat loss $Q_1 - Q_2$

$= 1956.5 - 1854.7$

$= 101.8w$

Step 6

Result

1. Heat loss $Q_1 = 1956.5w$

    $Q_2 = 1854.7\ w$

2. Reduction in heat loss $= 101.8w$

4.23. Problems based on shape Factor

1. Calculate the shape factor, F_{14}, for the arrangement shown in fig.

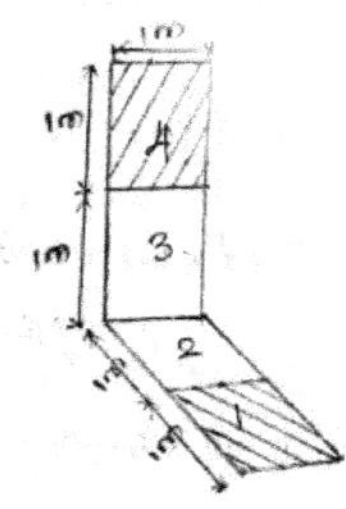

Step 1

Solution

Although charts of the shape factor for arrangement F_{14} are not available, it can be easily calculated by making use of figure which given below,

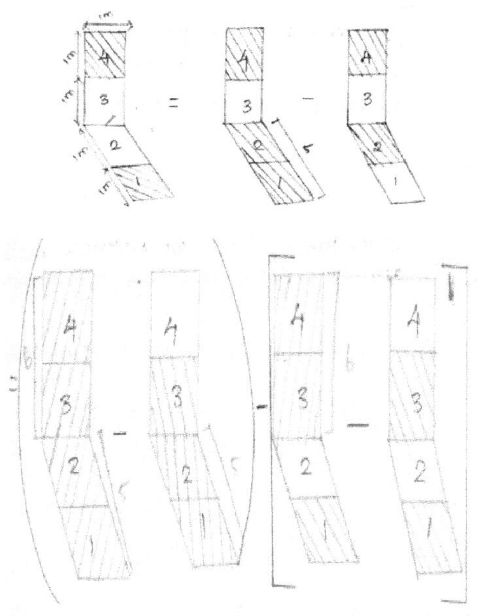

Step 2

$A_5 \quad = A_1 + A_2$

$A_6 \quad = A_3 + A_4$

$A_1 F_{14} = A_5 F_{54} - A_2 F_{24}$

$\qquad = A_5 [F_{56} - F_{53}) - A_2[F_{26} - F_{23}]$

$A_1 F_{14} = A_{1,2}\, F\,(1,2)\,(3,4) - A_{1,2}\,(F1,2)\,3 - A_2\,F_2\,(3,4) + A_2\,F_{23}$

$$F_{14} = \frac{A_5}{A_1}\left(F_{56} - F_{53}\right)$$

$\qquad = 2(0.15 - 0.11) - 1(0.24 - 0.2)$

$F_{14} = 0.04$

4.24. Problem based on Electrical Analogy

Two parallel plates of side 1m x 1m are spaced 0.5m apart are located in a very large room, the walls of which are maintained at a temperature of 27°C. One plate is maintained at a temperature of 900°C and the other at 400°C. Their emissivities are 0.2 and 0.5 respectively. If the plates exchange heat between themselves and surroundings, find the net heat transfer to each plate and to the room consider only the plate surfaces facing each others.

Step 1

Size of the plates = 1m x 1m

Distance between plates = 0.5m

Room temperature, T_3 = 27°C + 273

       = 300k

First plate temperature T_1 = 900°C + 273

       = 1173k

Second plate temperature T_2 = 400°C + 273

       = 673 k

Emissivity of first plate ε_1 = 0.2

Emissivity of second plate = 0.5

Step 2

To Find

i) Net heat transfer to each plate

ii) Net heat transfer to room.

Step 3

Solution

In this problem, heat exchange take place between two plates and the room. So this is three surface problem and the corresponding radiation network is given below.

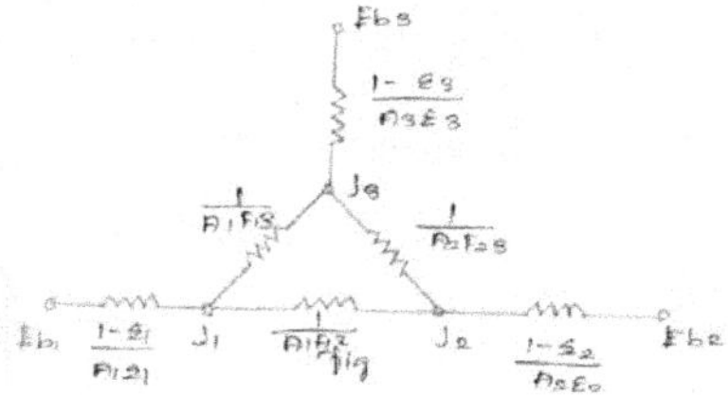

278

Area, $A_1 = 1 \times 1 = 1m^2$

$A_1 = A_2 = 1m^2$

Since the room is large, $A_3 = \infty$

Step 4

From electrical network diagram,

$$\frac{1-\varepsilon_1}{A_1\varepsilon_1} = \frac{1-0.2}{1 \times 0.2} = 4$$

$$\frac{1-\varepsilon_2}{A_2\varepsilon_2} = \frac{1-0.5}{1 \times 0.5} = 1 \qquad\qquad \left[\because A_3 = \infty\right]$$

$$\frac{1-\varepsilon_3}{A_3\varepsilon_3} = 0$$

Apply $\dfrac{1-\varepsilon_1}{A_1\varepsilon_1} = 4;\ \dfrac{1-\varepsilon_2}{A_2\varepsilon_2} = 1;\ \dfrac{1-\varepsilon_3}{A_3\varepsilon_3} = 0;$

Values in electrical network diagram.

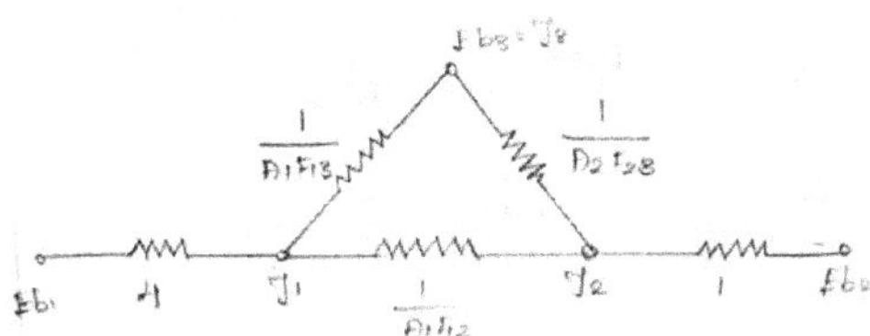

Step 5

To find shape factor F_{12}, refer HMT data book page no 91 &92 [Sixth Edition]

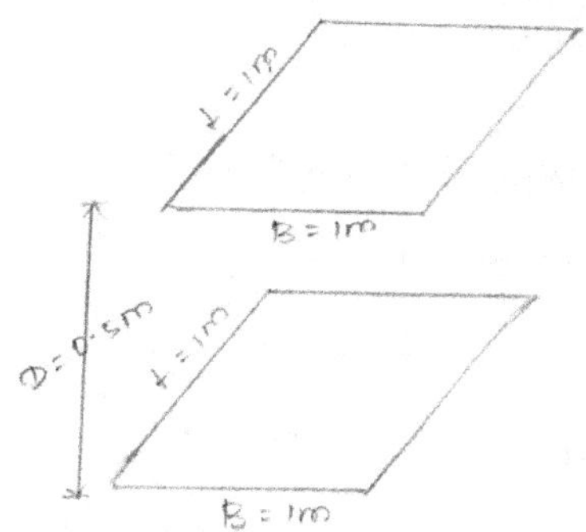

$$X = \frac{L}{D} = \frac{1}{0.5} = 2$$

$$Y = \frac{B}{D} = \frac{1}{0.5} = 2$$

Step 6

X values is 2, y value is 2, from that we can find corresponding shape factor value is 0.41525.[From table]

i.e., $F_{12} = 0.41525$

We know that,

$F_{11} + F_{12} + F_{13} = 1$

But, $F_{11} = 0$

$F_{13} = 1 - F_{12}$

$F_{13} = 1 - 0.41525$

$F_{13} = 0.5847$

Step 7

Similarly, $F_{21} + F_{22} + F_{23} = 1$

We know that, $F_{22} = 0$

$F_{23} \quad = 1 - F_{21}$

$= 1 - F_{12}$

$= 1 - 0.41525$

$F_{23} \quad = 0.5847$

From Electrical network diagram,

$$\frac{1}{A_1 F_{13}} = \frac{1}{1 \times 0.5847} = 1.7102$$

$$\frac{1}{A_1 F_{23}} = \frac{1}{1 \times 0.5847} = 1.7102$$

$$\frac{1}{A_1 F_{12}} = \frac{1}{1 \times 0.41525} = 2.408$$

Step 8

From Stefan – Boltxmann law'

$E_b \quad = \sigma T^4$

$Eb_1 = \sigma T^{.4}$

$$= 5.67 \times 10^{-8} \, [1173]^4$$

$$E_{b1} = 107.34 \times 10^3 \text{ w/m}^2$$

$$E_{b2} = \sigma T^4$$

$$= 5.67 \times 10^{-8} \, [673]^4$$

$$E_{b2} = 11.63 \times 10^3 \text{ w/m}^2$$

$$E_{b3} = \sigma T_4^{\,8}$$

$$= 5.67 \times 10^{-8} \, [300]^4$$

$$E_{b3} = 459.27 \text{ w/m}^2$$

From Electrical network diagram, we know that

$$E_{b3} = F_4 = 459.27 \text{ w/m}^2$$

Step 9

The radiosities J_1 and J_2 can be calculated by using kirchoff's law.

=> The sum of current entering the node J, is zero.

At Node T_1:

$$\frac{E_{b1} - J_1}{4} + \frac{J_2 - J_1}{\dfrac{1}{A_1 F_{12}}} + \frac{E_{b3} - J_1}{\dfrac{1}{A_1 F_{13}}} = 0$$

[From electrical network diagram]

$$\Rightarrow \frac{107.34 \times 10^3 - J_1}{4} + \frac{J_2 - J_1}{2.408} + \frac{459.27 - J_1}{1.7102} = 0$$

$$\Rightarrow 26835 - \frac{J_1}{4} + \frac{J_2}{2.408} - \frac{J_1}{2.408} + 268.54 - \frac{J_1}{1.7100} = 0$$

$$\Rightarrow 26835 - 0.25\,J_1 + 0.415\,J_2 - 0.415 J_1 + 268.54 - 0.5847 J_1 = 0$$

$$\Rightarrow -1.2497\,J_1 + 0.415\,J_2 = -27.10 \times 10^3$$

Step 10

$$\frac{J_1 - J_2}{\dfrac{1}{A_1 F_{12}}} + \frac{F_{b3} - J_2}{\dfrac{1}{A_2 F_{23}}} + \frac{E_{b2} - J_2}{2} = 0$$

$$\Rightarrow \frac{J_1 - J_2}{2.408} + \frac{459.27 - J_2}{1.7102} + \frac{11.63 \times 10^3 - J_2}{2} = 0$$

$$\Rightarrow \quad \frac{J_1}{2.408} - \frac{J_2}{2.408} + 268.54 - \frac{Je}{1.7102} + \frac{11.63 \times 10^3}{2} - \frac{J_2}{2} = 0$$

$$\Rightarrow \quad 0.415\,J - 0.415\,J_2 + 268.54 - 0.5847\,J_2 + 5.815 \times 10^3 - 0.5J_2 = 0$$

$$\Rightarrow \quad 0.415\,J_1 - 1.4997\,J_2 = -6.08 \times 10^3 \qquad \rightarrow (2)$$

Solving Equation (1) and (2)

$$-1.2497\,J_1 + 0.415\,J_2 = -27.10 \times 10^3$$

$$0.415\,J_1 - 1.4997\,J_2 = -6.08 \times 10_3$$

By solving, $J_2 = 11.06 \times 10^3 \ w/m^2$

$J_1 = 25.35 \times 10^3 \ w/m$

Heat lost by plate (1), $Q_1 = \dfrac{E_{b1} - J_1}{\dfrac{1 - \varepsilon_1}{A_1 \varepsilon_1}}$

[From electrical network diagram]

$$= \frac{107.34 \times 10^3 - 25.35 \times 10^3}{\dfrac{1 - 0.2}{1 \times 0.2}}$$

$Q_1 = 20.49 \times 10^3 \ w$

Step 11

Heat lost by plate (2), $Q_2 = \dfrac{E_{b2} - J_2}{\dfrac{1 - \varepsilon_2}{A_2 \varepsilon_2}} = \dfrac{11.63 \times 10^3 - 11.06 \times 10^3}{\dfrac{1 - 0.5}{1 \times 0.5}}$

$Q_2 = 570 \ w$

Step 12

Total heat lost by the plates (1) and (2) $Q = Q_1 + Q_2 = 20.49 \times 10^3 + 570$

$$Q = 21.06 \times 10^3 \ w$$

Total heat received or absorbed by the room $Q = \dfrac{J_1 - J_3}{\dfrac{1}{A_1 F_{13}}} + \dfrac{J_2 - J_3}{\dfrac{1}{A_2 F_{23}}}$

$$= \frac{25.35 \times 10^3 - 459.27}{1.7102} + \frac{11.06 \times 10^3 - 459.27}{1.7102}$$

$$\left[\because E_{b3} = J_3 = 459.27 \ ^{w}\!/_{m^2} \right]$$

$$Q = 20.752 \times 10^3 w$$

Step 13

Result

1. Net heat lost by each plates.

   $Q_1 = 20.49 \times 10^3$ w

   $Q_2 = 570$ w

2. Net heat transfer to the room

   $Q = 20.752 \times 10^3$ w

4.25. Formula Used in the Mean Beam Length

Description	Formula	
To find emissivity of Co_2 X-axis Curve Y-axis	From graph temperature (T) ink Pco2 XLmin (m-atm) 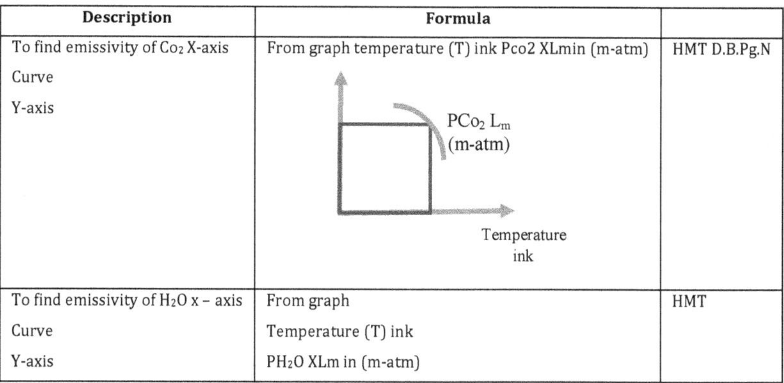	HMT D.B.Pg.N
To find emissivity of H_2O x – axis Curve Y-axis	From graph Temperature (T) ink PH_2O XLm in (m-atm)	HMT

4.26. Problems based on Mean Beam Length

1. A heating finance of total surface area $30m^2$ and volume $10m^3$ is maintained at a temperature of $1000°C$ over its entire volume. The total pressure of the combustion gases in 2 atm the partial pressure of water vapour is 0.1 atm and that of Co_2 is 0.3 atm. Calculate the emissivity of the gaseous mixture.

Step 1

Solution

The mean beam length for the gasoonsformula

$$Lms = 3.6\frac{V}{A} = 3.6 \times \frac{10}{30} - 1.2m$$

Now,

Pco_2, Lms = 0.3 x 1.2 = 0.36m atm

At $1000°C$ and PLms = 0.36

$\sum_{Co2} = 0.16$ at P = 1 atm

At P = 2 atm, PLms = 0.36m atm

$C_{co2} = 1.11$

$C_{co2}, \Sigma_{Co2} = 0.16 \times 1.11 = 0.1776$

Step 2

For H_2O, PH_2O

$L_{ms} = 0.1 \times 1.2 = 0.12$ matm

At 1000^oC and $P_{Lms} = 0.12$

$\varepsilon_{H_2O} = 0.12$

Now,

$$\frac{p+P}{2} = \frac{0.1+2}{2} = 1.05$$

At $P_{Lms} = 0.12$ and $\dfrac{p+P}{2} = 1.05$

$C_{H_2O} = 1.43$

$\varepsilon_{H_2O}, \varepsilon_{H_2O} = 1.43 \times 0.12 = 0.171$

Step 3

Since both Co_2 and H_2O are present in the gaseous mixture, another correction factor $\Delta\varepsilon$ can be found as follows:

$$\frac{\rho_{H_2O}}{(\rho_{H_2O} + \rho_{Co_2})} = \frac{0.1}{(0.1+0.3)} = 0.25$$

Step 4

$(P_{Co2} + Lms + P_{H_2O} + Lms) \quad = 0.36 + 0.12$

$$= 0.48m \text{ atm}$$

At $P_{Lms} = 0.48$ m atm and at abscissa of 0.25,

$\Delta\varepsilon = 0.035$

$\therefore$ Total emissivity of the gaseous mixture is

$\Sigma_r = \varepsilon_{CO_2} \cdot C_{CO_2} + \varepsilon_{H_2O}\, C_{H_2O} - \Delta\varepsilon$

$= 0.1776 + 0.171 - 0.035$

$= 0.3136$

UNIT 5

MASS TRANSFER

5.1. Mass Transfer

The process of transfer of Mass as a result of the species concentration difference in a mixture is known as Mass transfer.

Examples

1. Humidification of air in cooling tower.
2. Evaporation of petrol in the carburetter of an IC engine
3. The transfer of water vapour into dry air.
4. Dissolution of sugar added to a cup of coffee.

5.2. Modes of Mass Transfer

They are basically two modes of mass transfer.

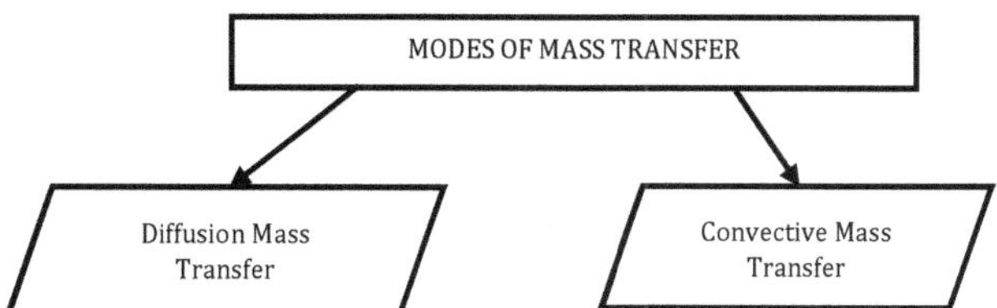

5.3. Diffusion Mass Transfer

It may be classified into two types,

1. Molecular diffusion
2. Eddy diffusion

5.4. Molecular Diffusion

- The transfer of water on a microscopic level as a result of diffusion from a region of higher concentration to a lower concentration in a mixture of liquids or gases is known as molecular diffusion.

5.5. Eddy Diffusion

- When one of the diffusion fluids is in turbulent motion, eddy diffusion takes place.
- Mass transfer is more rapid by eddy diffusion than by molecular diffusion.

5.6. Heat and Mass Transfer Analogy

S.NO	HEAT TRANSFER	MASS TRANSFER
1	The driving force for heat transfer is temperature gradient (ΔT).	The driving force for mass transfer is concentration gradient (Δc).
2	The heat transferred by conduction, convection and radiation.	The mass transfer from diffusion.
3	Thermal conductivity (k) : It is measure ability of heat conduct to transfer.	Mass diffusivity (C) : It is measure for how to commodity diffusion to medium.
4	Fourier law of heat transfer: The rate of heat conduction is proportional to the area measured normal to the direction of heat flow and to the temperature gradient in is direction $$Q = -KA \frac{dT}{dx}$$	Fick's law of mass transfer : The rate of mass conduction is proportional to the area measured normal to the direction of mass flow and to the concentration gradient in that direction. $$Q = -DA \frac{dC_A}{dy}$$
5	The rate of heat convection $Q_{conv} = h_{con}A\,(T_1 - T_\infty)$ hcon – heat transfer co-efficient	The rate of mass convection $Q_{conv} = h_{con}A\,(C_A - C_B)$ hcon – mass transfer coefficient
6	The heat convection is a heat transfer mechanism involve both heat conduction and fluid motion	The mass convection is a mass transfer mechanism involve both mass diffusion and bulk fluid motion.
7	In heat transfer, heat generated through out medium, convert some of nuclear energy to sensible thermal energy	Similarly, mass transfer some of energy convert through out medium.
8	**Prandyl number** $$P_r = \frac{c_p \cdot \mu}{k}$$ It significance, Pr = 1, velocity and thermal energy is coincidence almost equal each other.	**Schmit number** $$S_c = \frac{\mu}{\rho D} = \frac{U}{D}$$ It significance, Sc = 1, velocity and thermal energy is coincidence almost equal
9	**Lewis number** $$Le = \frac{\alpha}{D} = \frac{Sc}{Pr}$$ $$\frac{\text{Heat diffisivity}}{\text{mass diffisivity}}$$	**Lewis number,** $$Le = \frac{\alpha_m}{D_m}$$ $$\frac{\text{Heat diffisivity}}{\text{mass diffisivity}}$$
10	**Nusselt Number** $$Nu = \frac{hL}{K} = \frac{hd}{K}$$ Ratio of temperature gradient by conduction and convection at the surface	**Sherwood number** $$Sh = \frac{hm.L}{D} = \frac{hm.d}{D}$$ Ratio of concentration gradients at the boundary by diffusion and by convection
11	**Station Number** $$St = \frac{Nu}{Re.Pr}$$	**Station Number** $$St = \frac{Sh}{Re.Sc}$$

5.7. Convective Mass Transfer

Convective mass transfer is a process of mass transfer that will occur between a surface and a fluid medium when they are a different concentrations.

5.8. Concentrations

i. Mass Concentration (or) Mass Density

The mass concentration is defined as the mass of component per unit volume in a mixture. It is expressed in Kg/m^3.

$$\text{Mass Concentration} = \frac{\text{mass of a component}}{\text{Unit volume of mixture}}$$

ii. Molar Concentration (or) Molar Density

The ratio of number of molecules of component to that unit volume of mixture.

$$\text{Molar Concentration} = \frac{\text{Number of molecules of component}}{\text{Unit volume of mixture}}$$

iii. Mass Fraction

The ratio of mass concentration of species to the total mass density of the mixture.

$$\text{Mass fraction} = \frac{\text{Mass concentration of a species}}{\text{Total mass density}}$$

$$m_A = \frac{\rho A}{\rho}$$

iv. Mole Fraction

The ratio of mole concentration of a species to the total molar concentration

$$\text{Mole Fraction} = \frac{\text{Mole concentration of a species}}{\text{Total molar concentration}}$$

$$x_A = \frac{C_A}{C}$$

5.9. Fick's Law of Diffusion

- Consider a system as shown in fig.5.1. consider two gases are separately a and b.
- When the partition is removed, the two gases are diffuses through a one other until the equilibrium is established throughout of system.
- The diffusion rate is given by the fick law, that states molar flux in element per unit area is directly proportional to concentration gradient.

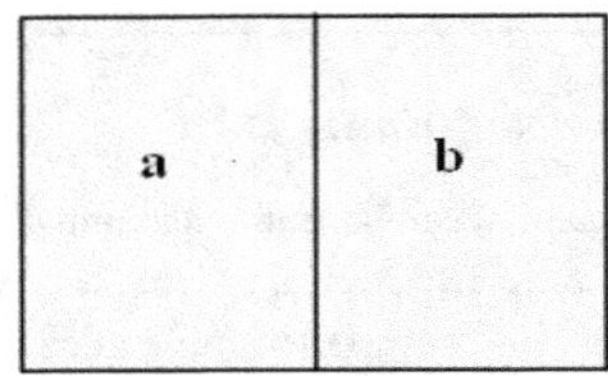

Fig. 5.1: Fick's Law of Diffusion

$$\rightarrow \quad \frac{M_a}{A} \propto \frac{d_{ca}}{d_x}$$

$$\rightarrow \quad \frac{M_a}{A} = -\,Dab.\frac{d_{ca}}{d_x}$$

$$\boxed{Na = \frac{M_a}{A} = -\,Dab.\frac{d_{ca}}{d_x}} \qquad \rightarrow (5.1)$$

- $Na = \dfrac{m_a}{A}$ molar flux in $\dfrac{kg-mole}{S-m^2}$
- Dab = Diffusion coefficient of species a and b in m²/s.
- d_{ca}/d_x = concentration gradient.

5.10. Steady State Diffusion Through a Plane Membrane

- Consider a plane membrane of thickness L, containing fluid 'a'. The concentrations of the fluid at the opposite wall faces are ca_1 and ca_2 respectively.
- Considering the diffusion in along x – axis then the controlling equation is,

$$\boxed{\frac{d^2ca}{dx^2} = 0} \qquad \rightarrow (5.2)$$

integrating the above equation

$$\boxed{\dfrac{dca}{dx} = c_1}$$

$$\rightarrow (5.3)$$

again integrating,

$$\boxed{C_a = C_1x + C_2}$$

$$\rightarrow (5.4)$$

Applying boundary conditions,

At, x = 0

$Ca_1 = Ca_2$

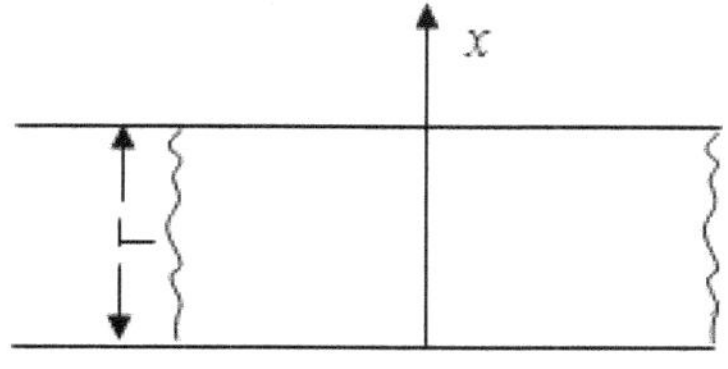

Fig 5.2

At, x = L

$Ca_2 = C_1L + C_2$ $[Ca_1 = C_2]$

$Ca_2 = C_1L + Ca_1$

$$C_1 = \dfrac{Ca_2 - Ca_1}{L} \qquad \rightarrow (5.5)$$

Sub C_1 and C_2 values in equation,

$$(5.4) \quad Ca = \left(\dfrac{Ca_2 - Ca_1}{L} \right) x + Ca_1$$

From fick's law, WKT.

$$\boxed{\text{Molar Flux, } \dfrac{ma}{A} = -Dab.\dfrac{dCa}{dx}}$$

$$\dfrac{ma}{A} = -Dab.\dfrac{d}{dx}\left(\dfrac{Ca_2 - Ca_1}{L} \right) x + Ca_1$$

$$\frac{ma}{A} = Dab \left(\frac{Ca_2 - Ca_1}{L} \right) \qquad \rightarrow (5.6)$$

$$\boxed{\text{Molar Flux, } \frac{ma}{A} = \frac{Dab}{L}[ca_2 - ca_1]}$$

$$\rightarrow (5.7)$$

Where,

- $\dfrac{ma}{A} \Rightarrow$ molar flux - $\dfrac{kg-mole}{S-m^2}$

- $Dab \Rightarrow$ Diffusion coefficient – m^2/S

- $Ca_1 \Rightarrow$ Concentration at inner side - $\dfrac{kg-mole}{m^3}$

- $Ca_2 \Rightarrow$ Concentration at outer side - $\dfrac{kg-mole}{m^3}$

- $L \Rightarrow$ thickness in m.

For spheres,

$L = r_2 - r_1$

$A = 4\,\pi r_1 r_2$

For cylinders,

$L = r_2 - r_1$

$$A = \frac{2\pi L(r_2 - r_1)}{l_n \left(\dfrac{r_2}{r_1} \right)}$$

Where,

r_1 = inner radius (m), r_2 = outer radius in (m).

5.11. Steady State Equimolar Counter Diffusion

- Consider two large chambers a and b connected by a passage as shown in fig.

- N_a and N_b are the steady state molar diffusion rates of components a and b respectively.

- Equimolar diffusion is defined as each molecules of 'a' is replaced by each molecule of 'b' of vice versa. The total pressure $p = P_a + P_b$ is uniform throughout the system.

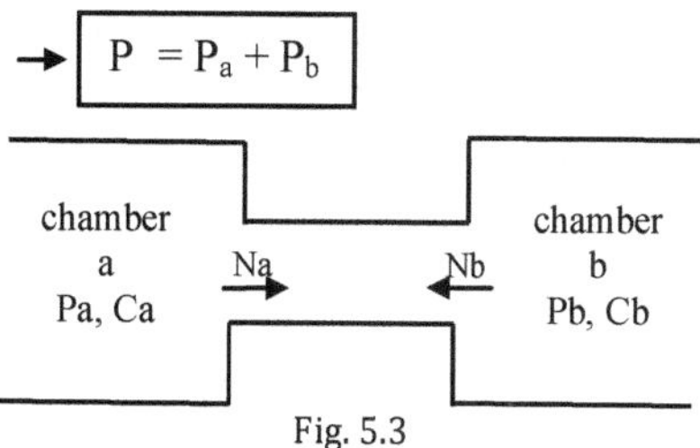

Fig. 5.3

Differentiating with respect to x,

$$\frac{dp}{dx} = \frac{dp_a}{dx} + \frac{dp_b}{dx}$$

$\rightarrow$(5.8)

Since, the total pressure of the system remains constant under steady state conditions,

$$\frac{dp}{dx} = \frac{dp_a}{dx} + \frac{dp_b}{dx} = 0$$

$$\frac{dp_a}{dx} = -\frac{dp_b}{dx}$$

$\rightarrow$(5.9)

Under steady state conditions, the total molar flux is zero,

$$Na + Nb = 0$$

$$Na = -Nb$$

$$-Dab.\frac{A}{GT}\frac{dp_a}{dx} = Dba\frac{A}{GT}\frac{dp_b}{dx}$$

$\rightarrow$ (5.10)

<u>From Fick's law,</u>

$$N_a = -Dab\,\frac{A}{GT}\frac{dp_a}{dx}$$

$$N_a = -D_{ba}\frac{A}{GT}\frac{dp_b}{d_x}$$

We know,

$$\frac{dp_b}{dx} = -\frac{dp_a}{dx} \text{ (from eqn 5.9)}$$

Sub in equation (5.10)

$$ -D_{ab}\frac{A}{GT}\frac{dp_a}{d_x} = -D_{b_a}\frac{A}{GT}\frac{dp_a}{d_x} \qquad * $$

$D_{ab} = D_{ba} = D$

So,

$$ Na = -D\frac{A}{GT}\frac{dp_a}{dx} $$

$$ Na = ma = -D\frac{A}{GT}\frac{dp_a}{dx} $$

$$ \frac{ma}{A} = \frac{-D}{GT}\frac{dp_a}{dx} \qquad * $$

integrating,

$$ N_a = \frac{m_a}{A} = \frac{-D}{GT}\int_1^2 \frac{dp_a}{dx} $$

$$ \text{Molar flux, } N_a = \frac{m_a}{A} = \frac{D}{GT}\left[\frac{Pa_1 - Pa_2}{x_2 - x_1}\right] \qquad \rightarrow(5.11) $$

Similarly,

$$ \text{Molar flux, } N_b = \frac{m_b}{A} = \frac{D}{GT}\left[\frac{Pb_1 - Pb_2}{x_2 - x_1}\right] \qquad \rightarrow(5.12) $$

Where,

- $\dfrac{ma}{A}$ = molar flux - $\left(\dfrac{kg-mole}{S-m^2}\right)$

- D = Diffusion Coefficient – (m^2/S)

- G = Universal constant

- A = Area – m^2

- T – Temperature – K

- Pa_1 & Pa_2 = Partial pressures at 1 & 2 in N/m^2.

5.12. Isothermal Evaporation of Water into Air

- Consider the iso thermal evaporation of water from a water surface.
- And it's diffusion through a stagnant air layer over it as shown in fig.5.4.
- The free surface of the water exposed to air in the tank.

For the Analysis of this Type of Mass Diffusion

1. System is isothermal and total pressure remains constant.
2. System is in steady state condition.
3. Air and water vapour behave as ideal gases.
4. There is slight air movement over the top of the tank to remove the water vapour

Which diffuses to that point.

From fick's law of diffusion, we can find,

$$\text{Molar flux, } \frac{ma}{A} = \frac{Dab}{GT} \frac{P}{(x_2 - x_1)} l_n \left(\frac{P - Pw_2}{P - Pw_1} \right)$$

$\rightarrow (5.13)$

Where,

- $\dfrac{ma}{A}$ = molar flux - $\dfrac{kg - mole}{S - m^2}$

- D = Diffusion Coefficient – m²/S

- T – Temperature – K

- P – Total pressure in bar

- Pw_1 & Pw_2 = Partial pressures of water vapour corresponding to saturation temperature and dry air in N/m².

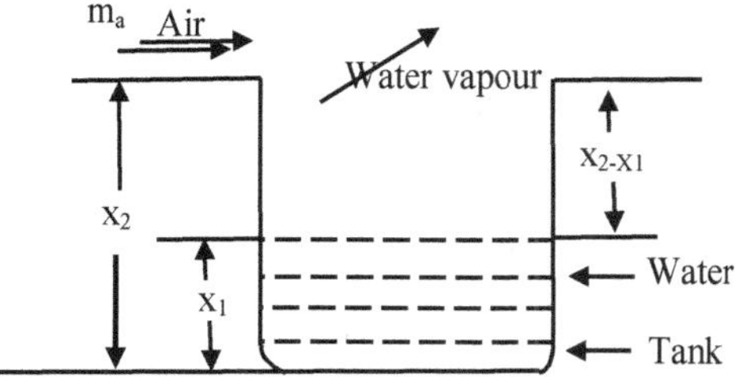

Fig. 5.4

Concentration		
1	Molar concentrations (C) (Or) molar density	$C = \dfrac{P}{GT}$ $(kg - mole /m^3)$ G – universal gas constant G = 8314 J/Kg – mole – k T – Temperature in k P – Partial pressure in N/m²
	Total concentration	$C = C1 + C2$
2	Mass densities (ρ) (Or) mass concentration	Molar concentration $C = \dfrac{\rho}{M}$ $\rho = C * M \left(\dfrac{kg}{m^3}\right)$
	Molecular weight	M_{D2} = 32 HMT DB Pg.No = 184 M_{N2} = 28
	Over all density	$\rho = \rho_1 + \rho_2$
3	Mass fraction (m)	$\dot{m}_1 = \dfrac{\rho_1}{\rho}$, $\dot{m}_2 = \dfrac{\rho_2}{\rho}$
4	Mole fractions (x)	$x_1 = \dfrac{C_1}{x}$, $x_2 = \dfrac{C_2}{c}$
5	Total pressure (p)	$Pv = mRT$ $P = \dfrac{m}{v}RT = \rho RT$ $\left(there\ fore\ R = \dfrac{G}{M}\right)$ Therefore G = 8314J/kg – mole –k Pressure in N/m²

5.14. Problems based on Concentration

1. A vessel contains a binary mixture of O_2 and N_2 with partial pressures in the ratio 0.21 and 0.79 at 15°C. If the total pressure of a mixture 1.1 bar calculate the following.

i. Molar concentrations

ii. Mass densities

iii. Mass fractions

iv. Molar fraction of each species.

Step 1

Given Data

* Partial pressure of O_2, P_{O_2} = 0.21 x Total pressure

    = 0.21 x 1.1 bar

    = 0.21 x 1.1 x 10^5 N/m²

* Partial pressure of N_2, P_{N_2} = 0.79 x Total pressure

    = 0.79 x 1.1 bar

    = 0.79 x 1.1 x 10^5 N/m²

* Temperature T = 15° + 273 = 288k

Step 2

To Find

1. Molar concentrations
2. Mass densities
3. Mass fractions
4. Molar fraction of each species

Step 3

WKT,

Molar concentration, $C = \dfrac{P}{GT}$

$$C_{O_2} = \dfrac{P_{O_2}}{GT}$$

$$= \dfrac{0.21 \times 1.1 \times 10^5}{8314 \times 288}$$

$$\boxed{C_{O_2} = 9.65 \times 10^{-3} \text{ kg} - \text{mole} /m^3}$$

$$C_{N2} = \dfrac{P_{N_2}}{GT} = \dfrac{0.79 \times 1.1 \times 10^5}{8314 \times 288}$$

$$\boxed{C_{N_2} = 0.03629 \text{ kg} - \text{mole} /m^3}$$

Step 4

Molar concentration $c = \rho/m$

$\rho = C \times m$

$\rho o_2 = Co_2 \times mo_2$

$\qquad = 9.65 \times 10^{-3} \times 32 \quad [\because \text{molecular weight } o_2 = 32]$

$\rho o_2 = 0.3088 \text{ kg/m}^3$

$\rho N_2 = CN_2 \times MN_2$

$\qquad = 0.03629 \times 28 \qquad\qquad \therefore [\text{ molecular weight } N_2 \text{ is } 28]$

$\qquad = 1.01612 \text{ kg/m}^3$

Overall density $\rho = \rho o_2 + \rho N_2 = 0.3088 + 1.01612$

$$\boxed{\rho = 1.3249 \text{ kg/m}^3}$$

Step 5

Mass Fractions

$$\dot{m}_{o2} = \frac{\rho o_2}{\rho} = \frac{0.3088}{1.3249}$$

$$\boxed{\dot{m}_{o2} = 0.233}$$

$$\dot{m}_{o2} = \frac{\rho N_2}{\rho} = \frac{1.01612}{1.3249}$$

wKT,

$$\boxed{\dot{m}_{o2} = 0.800}$$

Total concentrations $C = C_{O_2} + C_{N_2} = 9.65 \times 10^{-3} + 0.03629$

$$\boxed{C = 0.046}$$

$\qquad\qquad\qquad\qquad\qquad\qquad\qquad\qquad\qquad \text{kg} - \text{mole} /\text{m}^3$

Step 6

Mole Fractions

$$X_{O_2} = \frac{C_{O_2}}{C}$$

$$X_{O_2} = \frac{9.65 \times 10^{-3}}{0.046}$$

$$= 0.209$$

$$x_{N_2} = \frac{C_{N_2}}{C}$$

$$x_{N_2} = \frac{0.03629}{0.046}$$

$$= 0.789$$

Step 7

Result

1) $C_{O_2} = 9.65 \times 10^{-3}$ kg –mole/m³

    $C_{N_2} = 36.29 \times 10^{-3}$ kg-mole/m³

2) $\rho_{O_2} = 0.3088$ kg/m³

    $\rho_{N_2} = 1.0612$ kg/m³

3) $\dot{m}_{O_2} = 0.233$

    $\dot{m}_{N_2} = 0.800$

4) $X_{O_2} = 0.209$

    $X_{N_2} = 0.789$

Table 5.2

S. NO	Description	Plane membrane (wall)	Plane membrane (cylinder)	Equimolar counter diffusion	ISO thermal evaporating of water in to air
1	Diagram	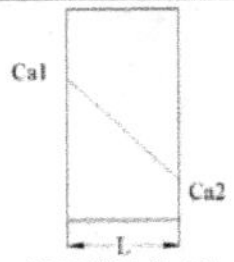 Ca1 ... Ca2 L HMT DB Pg.No 175	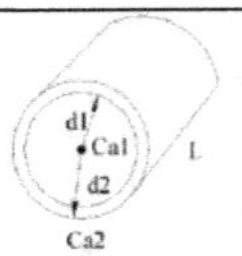 d1 · Ca1 / d2 / L Ca2 HMT DB Pg.No 176	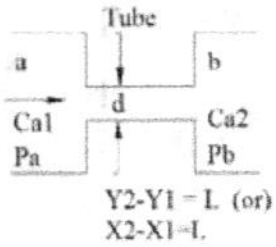 Tube a / d / b Ca1 / Ca2 Pa / Pb Y2-Y1 = L (or) X2-X1=L HMT DB Pg.No 175	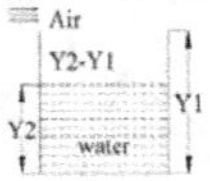 Air Y2-Y1 Y2 / water / Y1 HMT DB Pg.No 175
2	Diffusion flux (or) molar flux **Notes :** Molar concentration (C) Mass flux	$\dfrac{\dot{m}a}{A} = Dab\,\dfrac{C_{a1} - C_{a2}}{L}$ $\dot{m}_a$ - mass diffusivity rate . ($\dot{m}_a$ = kg – mole /s) ($\dfrac{\dot{m}a}{A}$ $\dfrac{Kg - mole}{S - m^2}$) L – Length in m D_{ab} - Diffusion co – efficient in m^2/s <u>On inner side</u> C_{a1}=solubility * inner pressure <u>On outer side</u> C_{a2}=solubility * External pressure If C_{a2} is not given , Take C_{a2} =0 Mass flux in Kg/S-m^2 = (molar flux * molecular weight) (Molecular weight HMT DB Pg.No – 184)	$\dot{m}a = \left(\dfrac{2\pi D_{ab}l(C_{a1} - C_{a2})}{\ln\left(\frac{r_2}{r_1}\right)}\right)$ r1 – inner radius in m r2 – outer radius in m l – Length in m	$\dfrac{\dot{m}a}{A} = \dfrac{D_{ab}}{RT} \cdot \dfrac{C_{a1} - C_{a2}}{y_2 - y_1}$ There fore $A = \dfrac{\pi}{4} * d^2 (m^2)$ Length of the pan L = $y_2 - y_1$ (cm) Universal R = 8314 J/kg – mole – K T – temperature (k) Notes : Mass rate of chamber = $\dot{m}a * molecular\ weight\ in\ \dfrac{kg}{s}$ (Molecular weight HMT DB Pg.No – 184)	$\dfrac{\dot{m}a}{A} = \dfrac{D_{ab}}{RT} \cdot \dfrac{P}{y_2 - y_1}$ There fore $A = \dfrac{\pi}{4} * d^2 (m^2)$ Notes : P = 1.01325 * 10^5N/m^2 Partial pressure of water rap our (P_{w1}) in N/m^2 Corresponding to saturation temperature (by using steam table) Partial pressure of dry air (P_{w2}) in N/m^2 (mostly P_{w2}=0 , because dry air) Identification Open pan (or) water rap our from the bottom of the test tube

5.15. Problem Based on Plane Membrane

1. Hydrogen gases at 3 bar and 1 bar are separated by a plastic membrane having thickness 0.25mm. The binary diffusion coefficient of hydrogen in the plastic is 9.1 x 10^{-8} m^2/s. The solubility of hydrogen in the membrane is 2.1 x 10^{-3} kg-mole/m^3 bar. An uniform temperature condition of 20º is assumed. Calculate the following,

    i.   molar concentration of hydrogen on both sides

    ii. molar flux of hydrogen

    iii. mass flux of hydrogen

Step 1

Given Data

Inside pressure P_1 = 3 bar,

Outside pressure P_2 = 1bar

Diffusion coefficient D_{ab} = 9.1 x 10^{-8} m^2/S

Solubility of hydrogen $= 2.1 \times 10^{-3} \dfrac{kg-mole}{m^{3}.bar}$

Temperature $T = 20°C$

Step 2

1. Molar concentrations on bot sides Ca_1 and Ca_2
2. Molar flux
3. Mass flux

Step 3

1. molar concentrations on inner side,

    Ca_1 = solubility x inner pressure

    $Ca_1 = 2.1 \times 10^{-3} \times 3$

    $Ca_1 = 6.3 \times 10^{-3} \dfrac{kg-mole}{m^{3}}$

    molar concentration at outer side

    Ca_2 = solubility x outer pressure

    $Ca_2 = 2.1 \times 10^{-3} \times 1$

    $Ca_2 = 2.1 \times 10^{-3} \dfrac{kg-mole}{m^{3}}$

Step 4

molar flux, $\dfrac{m_a}{A} = \dfrac{Dab}{L}[Ca_1 - Ca_2]$ {HMTDB P.No.175}

$$= \dfrac{9.1 \times 10^{-8}}{0.25 \times 10^{-3}}[6.3 \times 10^{-3} - 2.1 \times 10^{-3}]$$

$$\boxed{\dfrac{ma}{A} = 1.52 \times 10^{-6} \dfrac{kg-mole}{S-m^{2}}}$$

Step 5

Mass flux   = molar flux xmolecular weight

$$= 1.52 \times 10^{-6} \times 2 \quad (\dfrac{kg-mole}{S-m^{2}} \times (1/mole))$$

$$\boxed{\text{Mass flux} = 3.04 \times 10^{-6} \ kg/s\text{-}m^2}$$

Step 6

1) $Ca_1 = 6.3 \times 10^{-3} \dfrac{kg - mole}{m^3}$

2) $Ca_2 = 2.1 \times 10^{-3} \dfrac{kg - mole}{m^3}$

3) $\text{molar flux} = 1.52 \times 10^{-6} \dfrac{kg - mole}{S - m^2}$

4) $\text{mass flux} = 3.04 \times 10^{-6} \ kg/s\text{-}m^2$

2. Oxygen at 25°C and pressure of 2 bar is flowing through a rubber pipe of inside diameter 25mm and wall thickness 2.5mm. The diffusivity of O_2 through rubber is 0.21×10^{-9} m²/S and the solubility of O_2 in rubber is $3.12 \times 10^{-3} \dfrac{kg - mole}{m^3 - bar}$ find the loss of O_2 by diffusion per meter length in pipe.

Step 1

Given Data

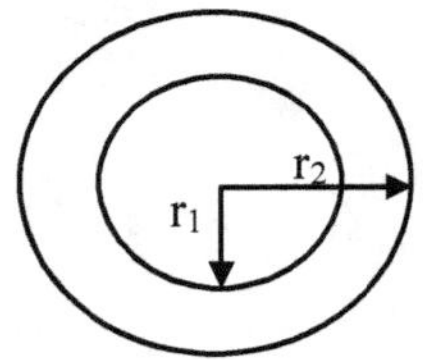

- Temperature,   T = 25°C
- Inside pressure, P_1 = 2 bar
- Inner diameter, d_1 = 25mm
- Inner radius, r_1 = 12.5mm = 0.0125m
- Thickness,   t = 2.5mm = 0.0025m
- Outer radius, r_2 = Inner radius + Thickness = 0.0125 + 0.0025

      r_2 = 0.015m

- Diffusion co-efficient = Dab = 0.21×10^{-9} m²/S
- Solubility = $3.12 \times 10^{-3} \dfrac{kg - mole}{m^3 - bar}$

Step 2

Loss of o_2 by diffusion per meter length.

Step 3

1. molar concentration on inner side,

Ca_1 = solubility x inner pressure

$Ca_1 = 3.12 \times 10^{-3} \times 2$

$Ca_1 = 6.24 \times 10^{-3} \dfrac{kg-mole}{m^3}$

molar concentration on outer side,

Ca_2 = solubility x outer pressure

$C_{q2} = 3.12 \times 10^{-3} \times 0$

$Ca_2 = 0$

[Assume the partial pressure o_2 and the outer surface of tube is zero]

wKT,

$$\boxed{\text{molar flux, } \frac{ma}{A} = \frac{Dab[Ca_1 - Ca_2]}{L}}$$

{HMTDB P.No.176}

Step 4

for cylinders,

$$L = r_2 - r_1; \quad A = \frac{2\pi L(r_2 - r_1)}{ln\left(\dfrac{r_2}{r_1}\right)}$$

$$m_a = \frac{2\pi L.\ Dab(Ca_1 - Ca_2)}{ln\left(\dfrac{r_2}{r_1}\right)}$$

$$m_a = \frac{2\pi \times 1 \times 0.21 \times 10^{-9} \times [6.24 \times 10^{-3}]}{ln\left(\dfrac{0.015}{0.0125}\right)}$$

$$\boxed{M_a = 4.51 \times 10^{-11} \dfrac{kg-mole}{S}}$$

(L = 1m)

Step 5

$$\text{Loss of oxygen} = 4.51 \times 10^{-11} \dfrac{kg-mole}{S}$$

5.16. Problem Based on Equimolar Counter Diffusion

Two large tanks, maintained at the same temperature and pressure are connected by a circular 0.15m diameter direct, which is 3m length. One tank contains a uniform mixture of 60 mole% ammonia and 40mole % air and the other tank contains a uniform mixture of 20mole % ammonia and 80 mole % air. The system is at 273k and 1.013 bar. Determine the rate of ammonia transfer between the two tanks. (Assuming a steady state mass transfer)

Step 1

Given Data

Diameter, d = 0.15m

length, $(x_2 - x_1)$ = 3m

$$Pa_1 = \frac{60}{100} = 0.6 \text{ bar} = 0.6 \times 10^5 \text{ N/m}^2$$

$$P_{b1} = \frac{40}{100} = 0.4 \text{ bar} = 0.4 \times 10^5 \text{ N/m}^2$$

$$Pa_2 = \frac{20}{100} = 0.2 \text{ bar} = 0.2 \times 10^5 \text{ N/m}^2$$

$$P_{b2} = \frac{80}{100} = 0.8 \text{ bar} = 0.8 \times 10^5 \text{ N/m}^2$$

T = 273K

P = 1.013 x 10⁵ N/m²

Step 2

a-ammonia

b – Air

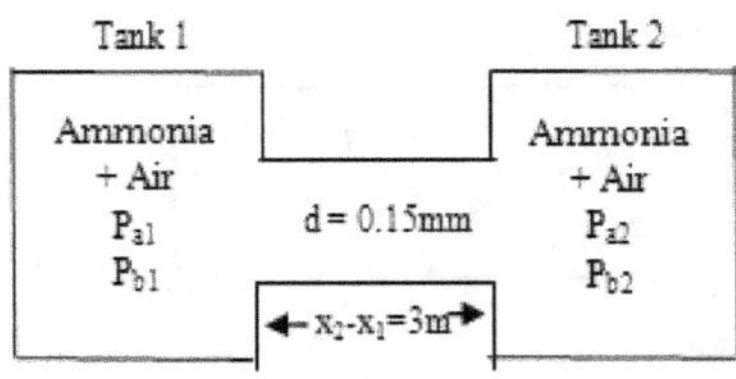

Step 3

Rate of ammonia transfer.

Step 4

WKT,

Equimolar counter Diffusion,

$$\text{Molar flux} = \frac{ma}{A} = \frac{Dab}{GT}\left[\frac{Pa_1 - Pa_2}{x_2 - x_1}\right]$$

{HMT DB P.No.175}

Where,

G – universal gas constant = 8314 J/kg.mole-k

A – Area - $\pi/4\ d^2 = \pi/4\ (0.15)^2$

A = 0.017m²

D_{ab} = Diffusion coefficient of ammonia and air.

$$Dab = 21.6 \times 10^{-6}\ m^2/S$$

$$\frac{ma}{0.017} = \frac{21.6 \times 10^{-6}}{8314 \times 273}\left[\frac{0.6 \times 10^5 - 0.2 \times 10^5}{3}\right]$$

$$\text{molar transfer rate of ammonia, ma} = 21.5 \times 10^{-9}\ \frac{kg-mole}{S}$$

Step 5

Mass transfer rate of ammonia = molar transfer rate of ammonia x molecular weight of ammonia

$= 21.5 \times 10^{-9} \times 17.03$

$$\text{Rate of ammonia transfer} = 3.66 \times 10^{-8}\ kg/S$$

Step 6

Mass transfer rate of ammonia $= 3.66 \times 10^{-8}\ kg/s$

5.17. Solved Problems on Isothermal Evaporation of Water into Air

1. Determine the diffusion rate of water from the bottom of a test tube of 25mm diameter and 35mm long into dry air at 25°C. Take diffusion co-efficient of water in air is 0.28 x 10⁻⁴ m²/S.

Step 1

Diameter, d = 25 mm = 0.025 m

Length(L)= $(x_2 - x_1)$ = 35mm = 0.035m

Temperature, T = 25°C + 273 = 298 K

Diffusion coefficient, Dab = 0.28 x 10^{-4} m^2/S .

Step 2

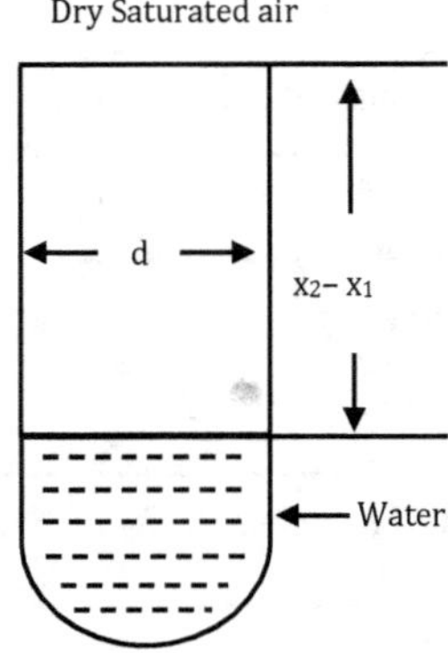

Step 3

To, find,

Diffusion rate of water.

Step 4

Isothermal Evaporation,

Molar flux, $$\dfrac{ma}{A} = \dfrac{Dab}{GT} \dfrac{P}{(x_2 - x_1)} \, ln \left(\dfrac{P - P_{w2}}{P - P_{w1}} \right) \qquad \rightarrow (1)$$

Where,

At = 25°C,

P_{w1} = Partial pressure at the bottom of the test tube corresponding to the saturation temperature.

At = 25°C, Condition,

P_{w1} = 0.03166 bar

(From R.S.Khurmi steam table, P.No.2)

$$P_{w1} = 0.03166 \times 10^5 \ \text{N}/\text{m}^2$$

Step 5

P_{w2} = Partial pressure at the top of the test tube. Here, air is dry and there is no water vapour, So, $P_{w2} = 0$

Area, $A = \dfrac{\pi}{4} d^2 = \dfrac{\pi}{4}(0.025)^2 = 4.90 \times 10^{-4} \text{m}^2$

Step 6

$$(1) \Rightarrow \frac{ma}{4.90 \times 10^{-4}} = \frac{0.28 \times 10^{-4}}{8314 \times 298} \times \frac{1.013 \times 10^5}{0.035} \times l_n\left[\frac{1.0131 \times 10^5 - 0}{1.013 \times 10^5 - 0.03166 \times 10^5}\right]$$

$$ma = 5.09 \times 10^{-10} \ \text{Kg} - \text{mole}/\text{S}$$

Step 7

WKT,

Mass rate of water vapour	=	Molar rate of water vapour	x	molecular weight of water vapour
		$= 5.09 \times 10^{-10} \times 18.016$		

[∴ molecular weight of water vapour = 18.016
HMT DB P.No. 183]

Mass transfer role of

water vapour $= 9.170 \times 10^{-9} \ \text{Kg}/\text{S}$

Step 8

Diffusion rate of water $= 9.170 \times 10^{-9} \ \text{Kg}/\text{S}$

2. A open pan 20cm in diameter and 8cm deep contains water at 25°C and   is exposed to dry atmospheric air. If the rate of diffusion of water vapour is 8.54 x 10⁻⁴ kg/h , Estimate the diffusion co-efficient of water in air.

Step 1

Given

Diameter, d	= 20cm = 0.20m
Length, $(x_2 - x_1)$	= 8cm = 0.08 m
Temperature, T	= 25°C + 273 = 298K
Diffusion rate	= 8.54 x 10⁻⁴ kg/h

$$= \frac{8.54}{3600} \times 10^{-4} = 2.37 \times 10^{-7} \; Kg/S$$

Step 2

To find: Diffusion coefficient, (D_{ab})

Step 3

WKT, molar rate of water vapour,   {HMTDB P.No. 175}

$$ma = \frac{ma}{A} = \frac{Dab}{GT} \frac{P}{(x_2 - x_1)} \ln\left(\frac{P - P_{w2}}{P - P_{w1}}\right) \qquad \rightarrow (1)$$

Mass rate of water vapour	=	Molar rate of water vapour	Molecular weight of steam

$$ma = \frac{Dab \times A}{GT} \times \frac{P}{(x_2 - x_1)} \times \ln\left(\frac{P - P_{w2}}{P - P_{w1}}\right) \times 18.016$$

$$A = \frac{\pi}{4}(d)^2 = \frac{\pi}{4}(0.20)^2 = 0.0314 m^2$$

G = universal gas constant = 8314 $J/kg-mole-k$

P = Total Pressure = 1atm = 1.013 bar = 1.013 x 10⁵ N/m^2

Step 4

P_{w1} = Partial pressure at the bottom of the test tube corresponding to saturation temperature

At 25°c.

$\Rightarrow$ P_{w1} = 0.03166 bar   (or)

P_{w1} = 0.03166 x 10^5 N/m^2

$\Rightarrow$ P_{w2} = 0 {$\therefore$ air is dry, no water vapour}

Step 5

$$(1) \Rightarrow 2.37 \times 10^{-7} = \frac{D_{ab} \times 0.0314}{8314 \times 298} \times \frac{1.013 \times 10^5}{0.08} \times \ln\left[\frac{1.013 \times 10^5 - 0}{1.013 \times 10^5 - 0.03166 \times 10^5}\right] \times 18.016$$

$$D_{ab} = 2.58 \times 10^{-5} \; \frac{m^2}{S}$$

Step 6

Diffusion Co-efficient, $D_{ab} = 2.58 \times 10^{-5} \; \dfrac{m^2}{S}$

5.18. Convective Mass Transfer

Convective mass transfer is a process of mass transfer will occur between a surface and a fluid medium. When they are at different concentrations.

5.19. Types of Convective Mass Transfer

1.   Free convective mass transfer
2.   Forced Convective mass transfer.

5.20. Free Convective Mass Transfer

If the fluid motion is produced due to change in density resulting from concentration gradients the made of mass transfer is free convective mass transfer.

Ex: Evaporation of Alcohol.

5.21. Forced Convective Mass Transfer

If the fluid motion is artificially created by means of an external force this type of mass transfer is known as forced convective mass transfer.

Ex: The evaporation of water from an ocean when air blows over it.

5.22. Dimensionless Groups

1. Reynolds Number (Re)

The ratio of inertia force to that viscous force

$$Re = \frac{Inertia\ force}{Viscous\ force} = \frac{ux}{v}$$

If a flat plate,

if $Re < 5 \times 10^5$, - Flow is laminar

     $Re > 5 \times 10^5$, ~ flow is turbulent.

2. Schmidt Number (SC)

The ratio of molecular diffusivity of momentum to that molecular diffusivity

$$Sc = \frac{molecular\ diffusivity\ of\ momentum}{molecular\ diffusivity\ of\ mass}$$

$$Sc = \frac{v}{D_{ab}}$$

3. Scherwood Number (S_h)

The ratio of concentration gradients at the boundary

$$S_h = \frac{h_{m.x}}{D_{ab}}$$

Where,

h_m – mass transfer coefficient – m/s

D_{ab} – Diffusion coefficient – m^2/s

x – length – m.

5.23. Formula Used in Flat Plate

		CONVECTIVE MASS TRANSFER				
S.NO	DESCRIPTION	FORCED CONVECTIVE PLATE	HMT DB Pg.N o	FORCED CONVECTIVE PIPE/CYLINDER /INTERNAL FLOW	HMT DB Pg.No
1	Velocity	Velocity is given		Velocity is given	
2	Properties	Properties of air ,from fluid temperature $(T\infty)$ The water For air Kinematic viscosity (V) m²/s	22 34	Properties of air ,from fluid temperature $(T\infty)$ The water For air Kinematic viscosity (V) m²/s	22 34
3	Reynolds number (Re)	$$Re = \frac{U.x}{v}\ (or)\ \frac{U.L}{v}$$ i) Laminar flow $Re < 5 \times 10^5$ ii) Turbulent flow $Re > 5 \times 10^5$	112 117	$$Re = \frac{UD}{V}$$ i) Laminar flow Re <2000 ii) Turbulent flow Re > 2000	112
4	Schmidt number (Sc)	$$Sc = V/D_{ab}\ \text{(or)}$$ $\mu/\rho .D_{ab}$ D_{ab} = diffusion of co-efficient	112	$$Sc = V/D_{ab}$$	112
5	Scher wood number (Sh)	i) For Laminar flow $Sh = 0.664\ (Re)^{0.5}(Sc)^{0.333}$ ii) For turbulent flow from leading edge $Sh = 0.0296\ (Re)^{0.8}(Sc)^{0.333}$ iii) For laminar – turbulent flow(combined flow) $Sh = [0.037Re^{0.8} - 871]SC^{0.333}$	176 177 178	i) For laminar flow $Sh = 3.66$ ii) For turbulent flow $Sh = 0.023(Re)^{0.83}(SC)^{0.44}$	177
6	To find (hm) mass transfer co-efficient in m/s	$$sh = \frac{h_m.L}{D_{ab}}$$ $h_m = \left(sh * \frac{D_{ab}}{L}\right) l_n \frac{m}{s}$	112	$$sh = \frac{h_m.d}{D_{ab}}$$ $h_m = \left(sh * \frac{D_{ab}}{d}\right) l_n \frac{m}{s}$	112

5.24. Problem Based On Flat Plate

1. Dry air at 20°C [$\rho = 1.2$ kg/m^3 , $V = 15 \times 10^{-6}$ m^2/s , $D_{as} = 4.2 \times 10^{-5}$ m^2/s] flows over a flat plate of length 50 cm, which is covered with a thin layer of water a velocity of 1 m/s. estimate the local mass transfer coefficient at a distance of 10cm from the leading edge and the average mass transfer coefficient.

Step 1

- Fluid temperature, $\qquad T_\infty = 20°C$

- Density, $\qquad \rho = 1.2$ kg/m^3

- Kinematic visoasity, $\qquad v = 15 \times 10^{-6}$ m^2/s

- Diffusion Co-efficient $\quad Dab = 4.2 \times 10^{-5}$ m^2/s

- Length $\qquad L = 50cm = 0.50$ m

- Velocity $\qquad U = 1$ m/s

- Distance $\qquad x = 10$ cm $= 0.10$m.

Step 2

1. Local mass transfer coefficient, h_x at a distance of 0.10m.
2. Average mass transfer coefficient, h_m for entire length.

Step 3

Case (i): Local mass transfer coefficient at x = 0.10m

WKT,

Reynolds Number, $R_e = \dfrac{Ux}{V}$ $\qquad\qquad\qquad$ {HMT DB P.NO. 112}

$$= \frac{1 \times 0.1}{15 \times 10^{-6}}$$

R_e $\quad = 6666.67 < 5 \times 10^5$

Since Re $< 5 \times 10^5$, Flow is laminar

Step 4

For laminar flow flat plate:

Local Sherwood number,

(HMTDB P.No. 176)

$$S_h = 0.332 \, (Re)^{0.5} \, (SC)^{0.333} \qquad \rightarrow (1)$$

$$S_c = \text{Schmidt number} = \frac{V}{D_{ab}}$$

$$S_c = \frac{15 \times 10^{-6}}{4.2 \times 10^{-5}} = 0.357$$

Sub S_C, R_e values in eqn (1)

$$1 \Rightarrow S_h = 0.332 \, (6666.67)^{0.5} \, (0.357)^{0.333}$$

$$S_h = 19.24$$

WKT,

Sherwood number,

$$\boxed{S_h = \frac{h_x \times x}{D_{ab}}}$$

[HMTDB P.No.112]

$$19.24 = \frac{h_x \times 0.1}{4.2 \times 10^{-5}}$$

✴ $h_x = 8.08 \times 10^{-3} \text{ m/s}$

Step 5

Case (ii): Average mass transfer co-efficient h_m, for entire length.

WKT,

$$\boxed{\text{Reynolds number, Re} = \frac{UL}{V}}$$

{HMT DB P.No. 112}

$$= \frac{1 \times 0.50}{15 \times 10^{-6}}$$

$Re = 3.33 \times 10^4 < 5 \times 10^5$

Since $Re < 5 \times 10^5$, flow is laminar

Step 6

For Flat plate laminar flow:

$$S_h = 0.664 \, (Re)^{0.5} \, (Sc)^{0.333}$$

(HMTDB P.No. 176)

$$S_h = 0.664 \, (3.33 \times 10^4)^{0.5} \, (0.357)^{0.333}$$

$$\boxed{S_h = 85.99}$$

WKT,

$$S_h = \frac{h_m \, x L}{D_{ab}} \qquad \{HMT\ DB\ P.No.\ 112\}$$

$$85.99 = \frac{h_m \times 0.50}{4.2 \times 10^{-5}}$$

✳ $h_m = 0.007$ m/s

Result

1. Local mass transfer coefficient

$$h_x = 8.08 \times 10^{-3} \ \text{m}/\text{s}$$

2. Average mass transfer coefficient for entire length $(h_m) = 0.007 \ \text{m}/\text{s}$.

5.25. Problems Based on Internal Flow

[Pipes and Cylinders]

1. Air at 1 atm and 25°C containing small quantities of iodine flow with a velocity 6.2 m/s inside a 35mm diameter tube, calculate the mass transfer coefficient for iodine. The thermo physical properties of air are,

$$v = 15.5 \times 10^{-6} \ \text{m}^2/\text{s}$$

$$D_{ab} = 0.82 \times 10^{-5} \ \text{m}^2/\text{s}$$

Step 1

Given Data

Pressure, p = 1 atm = 1.013 bar

Fluid temperature, T_∞ = 25°C

Velocity, U = 6.2 m/s

Diameter, D = 35mm = 0.035 m.

Kinematic Viscosity, v = 15.5 x 10^{-6} m^2/s

Diffusion Coefficient, D_{ab} = 0.82 x 10^{-5} m^2/s .

Step 2

To Find

Mass transfer coefficient, h_m= ?

Step 3

WKT,

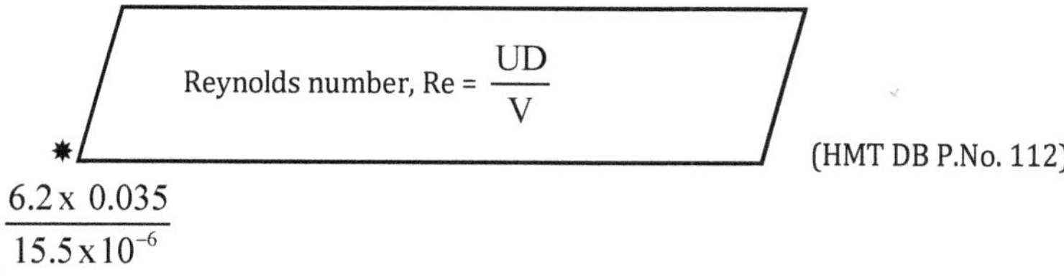

$$\text{Reynolds number, Re} = \frac{UD}{V}$$

(HMT DB P.No. 112)

$$= \frac{6.2 \times 0.035}{15.5 \times 10^{-6}}$$

Re = 14000 > 2000

(Since Re > 2000, Flow is turbulent.)

Step 4

✳ Scherwood number (S_h) = 0.023 x $(Re)^{0.83}$ x $(Sc)^{0.44}$

(HMTDB P.No. 177)

Where,

Sc – Schmidt number = $\dfrac{V}{Dab}$

(HMT DB P.No. 112)

$$Sc = \frac{15.5 \times 10^{-6}}{0.82 \times 10^{-5}}$$

$Sc = 1.809$

Sub, Sc& Re values in eqn (1)

$(1) \Rightarrow S_h = 0.023 \times (14000)^{0.83} \times (1.890)^{0.44}$

$S_h = 84.07$

Step 5

WKT,

Sherwood number, $S_h = \dfrac{h_m . D}{Dab}$ {HMTDB P.No. 112}

$$84.07 = \frac{h_m \times 0.035}{0.82 \times 10^{-5}}$$

$h_m = 0.0196$ m/s

Step 6

Result

Mass transfer coefficient, $h_m = 0.0196$ m/s.

<h1 align="center">Two Marks Questions and Answers</h1>

<h2 align="center">Unit 1: Conduction</h2>

1. State Fourier's Law of Heat Conduction.

The rate of heat conduction is proportional to the area measured normal to the direction of heat flow and to the temperature gradient in that direction.

$Q \propto -A \, (dT/dx)$

$Q = -KA \, (dT/dx)$

Where A-Area in m^2

(dT/dx)- Temperature gradient in K/m

K-Thermal conductivity in W/mK

2. Define thermal conductivity and list out the factors affecting the thermal conductivity.

Thermal Conductivity

It is defined as the ability of a substance to conduct the heat.

Factors Affecting the Thermal Conductivity

(i) Moisture

(ii) Density of material

(iii) Pressure

(iv) Temperature

(v) Structure of material

3. State Newton's law of cooling or State Newton's law of convection.

Heat transfer by convection is given by Newton's law of cooling: **$Q = hA \, (T_s - T_\infty)$**

Where, A -Area exposed to heat transfer in m^2

H - Heat transfer coefficient in $W/m^2 K$

T_s - Temperature of the surface in K

T_∞ - Temperature of the fluid in K

4. Define Overall heat transfer coefficient.

The overall heat transfer by combined mode is usually expressed in terms of an overall conductance or overall heat transfer coefficient "U"

$Q = UA \, \Delta T$

Q- Heat transfer in Watts

U- Overall heat transfer coefficient in $W/m^2 K$

ΔT – Temperature difference in K

5. **What are the modes of heat transfer?**

(i) Conduction

Heat transfer within the same medium (Solid to Solid)

Ex: Heating a steel rod

(ii) Convection

Heat transfer from one medium to another medium (Solid to liquid)

Ex: Heating of a pan which is full of water

(iii) Radiation

Heat transfer from one medium to another without any transmitting medium.

Ex: Sun light direct to the earth

6. **What is critical radius of insulation or critical thickness?**

Addition of insulating material on a surface does not reduce the amount of heat transfer rate always. In fact under certain circumstances it actually increases the heat loss up to certain thickness of insulation. "The radius of insulation for which the heat transfer is maximum is called critical radius of insulation" and the corresponding thickness is called "critical thickness".

7. **Define fins or extended surfaces.**

It is impossible to increase the heat transfer rate by increasing the surface of heat transfer.

The surfaces used for increasing heat transfer are called "Extended surfaces" or sometimes known as "fins".

8. **State the applications of fins.**

(i) Cooling of electronic components

(ii) Cooling of motor cycle engines

(iii) Cooling of transformers

(iv) Cooling of small capacity compressors

9. **Define fin efficiency and fin effectiveness.**

Fin Efficiency

It is the ratio of actual heat transferred to the maximum possible heat transferred by the fin.

$$\eta_{fin} = Q_{fin} / Q_{max}$$

Fin Effectiveness

It is the ratio of heat transfer with fin to the heat transfer without fin.

$$\epsilon_{fin} = Q_{with\ fin} / Q_{without\ fin}$$

10. Differentiate steady state and unsteady (Transient) heat conduction.

Steady State Heat Conduction

If the temperature of a body does not vary with time, it is said to be in a steady state and that type of conduction is known as "steady state heat conduction".

Unsteady State Heat Conduction

If the temperature of a body varies with time, it is said to be in a Unsteady state and that type of conduction is known as "Unsteady state heat conduction or transient heat conduction".

11. What is meant by lumped heat analysis?

In a Newtonian heating or cooling process the temperature throughout the solid is considered to be uniform at a given time.

Such an analysis is called "Lumped heat analysis".

Here Bi always less than 1. (Bi < 1)

12. What is meant by semi-infinite solid?

In a semi-infinite solid, at any instant of time, there is always a point where the effect of heating or cooling at one of its boundaries is not felt at all.

At this point the temperature remains unchanged.

The biot number value is ∞

13. What is meant by infinite solid?

A solid which extends itself infinitely in all directions of space is known as infinite solid.

Here the biot number value is in between 0.1 and 100.

i.e, $0.1 < Bi < 100$

14. Define Biot number and mention its significance.

It is defined as the ratio of the internal conductive resistance to the surface convective resistance.

$$Bi = hL_c / K$$

Significance

It is used to find Lumped heat analysis, Semi-infinite solids and infinite solids.

15. What are Heisler chart.

In Heisler chart, the solutions for the temperature distributions and heat flows in a plane walls, long cylinders and spheres with finite internal and surface resistance are presented.

It is the analytical solutions in the form of graphs.

Unit 2: Convection

Part-A (2 Marks with Answers)

1. What is meant by free or natural convection? Give examples.

If the fluid motion is produced due to change in density resulting from temperature gradients, the mode of heat transfer is said to be free or natural convection.

Ex:

(i) Ice Melting (Heat move to the ice from air-This causes the melting from a solid to liquid)

(ii) Steaming cup of hot tea. (The steam is showing heat being transferred in the air)

2. What is forced convection? Give examples.

If the fluid motion is artificially created by means of an external force like a blower or fan, that type of heat transfer is known as forced convection.

Ex: Steam turbines, Air conditioning, Heat exchangers, Car radiator using fluid

3. What are the dimensionless parameters used in forced convection.

(i)   Reynolds number (Re)

(ii) Nusselt number (Nu)

(iii) Prandtl number (Pr)

4. What are the dimensionless parameters used in free convection.

(i)   Grashof number (Gr)

(ii) Nusselt number (Nu)

(iii) Prandtl number (Pr)

5. Indicate the concept or significance of boundary layer.

1. A thin region near the body called the boundary layer where the velocity and the temperature gradients are large.

2. The region outside the boundary layer where the velocity and the temperature gradients are very nearly equal to their free stream values.

6. Define critical Reynolds number. What is the typical value for flow over a flat plate and flow through a pipe?

The Reynolds number at which flow changes from laminar to turbulent is called critical Reynolds number.

For plate, it is $5 * 10^5$

For Pipe it is 2300.

7. What is hydrodynamic and thermal boundary layer?

The velocity of the fluid is less than 99% of free stream velocity, it is said to be "hydro dynamic boundary layer".

The temperature of the fluid is less than 99% of free stream temperature, it is said to be "thermal boundary layer".

8. Define Reynolds number & Prandtl number.

Reynolds number (Re) is defined as the ratio of inertia force to the viscous force.

Prandtl number (Pr) is defined as the ratio of the momentum diffusivity to the thermal diffusivity.

9. Define Nussselt number & Groshof number.

Nusselt number (Nu) is defined as the ration of the heat flow by convection process under a unit temperature gradient to the heat flow rate by conduction under a unit temperature gradient through a stationary thickness (L) of mater.

Groshof number (Gr) is defined as the ratio of product of inertia force and buyoncy force to the square of the viscous force.

10. How does laminar flow differ from turbulent flow?

Laminar Flow

Laminar flow sometimes called stream line flow.

In this type of flow, the fluid moves in a layers and each fluid particle follows a smooth continuous path.

The fluid particles in each layer remain in orderly sequence without mixing with each other.

Turbulent Flow

In addition to the laminar type of flow a distinct irregular flow is frequently observed in nature. This type of flow is called turbulent flow. The path of any individual particle is zig-zag and irregular.

Part-A (2 Marks with Answers)

1. Define boiling and condensation.

Boiling

The change of phase from liquid to vapor state is known as boiling.

Condensation

The change of phase from vapor to liquid state is known as Condensation.

2. Give the application of boiling and condensation.

(i) Thermal and nuclear power plant

(ii) Refrigeration systems

(iii) Air conditioning systems

(iv) Process of heating and cooling

3. What is meant by pool boiling?

If heat is added to a liquid from a submerged solid surface, the boiling process is referred to as pool boiling. In this case the liquid above the hot surface is essentially stagnant and its motion near the surface is due to free convection and mixing induced by bubble growth and detachment.

4. What is meant by film wise condensation?

The liquid condensate wets the solid surface, spreads out and forms a continuous film over the entire surface is known as film wise condensation.

5. What is meant by drop wise condensation? Give its merits.

The vapor condensation, the vapor condenses into small liquid droplets of various sizes which fall down the surface in a random fashion.

Merits

A large portion of the area of the plate is directly exposed to vapor. The heat transfer rate in drop wise condensation is 10 times higher than in film condensation.

6. What is heat exchanger? What are the types of heat exchangers?

A heat exchanger is defined as equipment which transfers the heat from a hot fluid to cold fluid.

Types

(i) Direct contact heat exchangers

(ii) Direct contact heat exchangers

(iii) Surface heat exchangers

(iv) Parallel flow heat exchangers

(v) Counter flow heat exchangers

(vi) Cross flow heat exchangers

(vii) Shell and tube heat exchangers

(viii) Compact heat exchangers

7.   What do you meant by parallel and counter flow heat exchangers?

Parallel Flow Heat Exchangers

In this type of heat exchanger, hot and cold fluids move in the same direction.

Counter Flow Heat Exchangers

In this type of heat exchanger, hot and cold fluids move in parallel but opposite directions.

8.   Write short notes on compact heat exchangers.

There are many special purpose heat exchangers called compact heat exchangers. They are generally employed when convective heat transfer coefficient associated with one of the fluids is much smaller than that associated with the other fluid.

9.   What is meant by LMTD?

We know that the temperature difference between the hot and cold fluids in the heat exchanger varies from point to point. In addition various modes of heat transfer are involved.

Therefore based on the concept of appropriate mean temperature difference, also called logarithmic mean temperature difference, the total heat transfer rate in the heat exchanger is expressed as

$$Q = UA\ \Delta T_m$$

Where

U   - Overall heat transfer coefficient in W/m^2K, A- Area in m^2, ΔT_m - LMTD

10. What is meant by fouling factor?

We know, the surfaces of heat exchangers do not remain clean after it has been in use for some time. The surfaces become fouled with scaling or deposits. The effect of these deposits affecting the value of overall heat transfer coefficient. This effect is taken care of by introducing an additional thermal resistance called the fouling resistance.

11. What is meant by heat exchanger effectiveness?

It is defined as the ratio of actual heat transfer to the maximum possible heat transfer.

$$Effectiveness = Q/Q_{max}$$

12. What is meant by NTU? Give its expression.

Number of Transfer Unit (NTU) is a method which is used to determine the inlet or exit temperatures of heat exchangers.

$$NTU = UA/C_{min}$$

Unit 4: Radiation

Part-A (2 Marks with Answers)

1. Define emissive power.

It is defined as the total amount of radiation emitted by a body per unit time per unit area.

It is denoted as E_b and it is expressed in W/m^2.

2. Define monochromatic emissive power.

The energy emitted by the surface at a given length per unit time per unit area in all directions is known as monochromatic emissive power.

3. What is meant by absorptivity, reflectivity and transmissivity?

Absorptivity

It is the ratio between radiations absorbed to the incident radiation.

Reflectivity

It is the ratio between radiations reflected to the incident radiation.

Transmissivity

It is the ratio between radiations transmitted to the incident radiation.

4. What is black body?

If a body absorbs all incident radiation, regardless of wavelength and direction, then it is said to be black body.

For a prescribed temperature and wavelength, no surface can emit more energy than black body.

5. State Planck's Distribution law.

It states that the relationship between the monochromatic emissive power of a black body and wavelength of a radiation at a particular temperature.

6. Wien's Displacement law.

It states that the relationship between temperature and wavelength corresponding to the maximum spectral emissive power of the black body at that temperature.

7. State Stefan-Boltzmann law.

It states that the emissive power of a black body is proportional to the fourth power of absolute temperature.

$$E_b \propto T^4$$
$$E_b = \sigma T^4$$

Where,

E_b - Emissive power in W/m^2

σ – Stefan Boltzmann-constant=$5.67*10^{-8}$ W/m^2K^4

T – Temperature in K

8. Define emissivity.

It is defined as the ability of the surface of a body to radiate heat.

It is also defined as the ratio of emissive power of a body to the emissive power of a black body at equal temperature.

9. What is meant by gray body?

If a body absorbs a definite percentage of incident radiation irrespective of their wavelength, the body is known as gray body.

The emissive power of a gray body is always less than that of black body.

10. State Kirchhoff's law of radiation.

It states that the ratio of total emissive power to the absorptivity is constant for all surfaces which are in thermal equilibrium with the surroundings.

$$E1/\alpha1 = E2/\alpha2 = E3/\alpha3$$

11. Define intensity of radiation.

It is defined as the rate of energy leaving a space in a given direction per unit solid angle per unit area of the emitting surface normal to the mean direction in space.

$$I_n = E_b / \pi$$

12. State Lambert's cosine law.

It states that the total emissive power from a radiating plane surface in any direction proportional to the cosine of the angle of emission.

$$E_b \propto \cos \theta$$

13. What is the purpose of radiation shields?

Radiation shields are constructed from low emissivity (high reflectivity) materials.

It is used to reduce the net radiation transfer between two surfaces.

14. Differentiate irradiation and radiosity.

Irradiation

It is defined as the total radiation incident upon a surface per unit time per unit area.

It is expressed in W/m^2

Radiosity

It is used to indicate the total radiation leaving a surface per unit time per unit area. It is expressed in W/m^2

15. What are the assumptions made to calculate radiation exchange between the surfaces?

a. All surfaces are considered to be either black or gray.

b. Radiation and reflection process are assumed to be diffuse.

c. The absorptivity of a surface is taken equal to its emissivity and independent of temperature of the source of the incident radiation.

16. Discuss the radiation characteristics of carbon dioxide and water vapor.

The CO_2 and H_2O both absorb and emit radiation over certain wavelength regions called absorption bands.

The radiation in these gases is in a volume phenomenon.

The emissivity of CO_2 and the emissivity of H_2O at a particular temperature increases with partial pressure and mean beam length.

17. What is meant by shape factor and mention its significance.

It is defined as the fraction of the radiate energy that is diffused from one surface element and strikes the other surface directly with no intervention reflections.

It is used to analyze the radiative heat exchange between two surfaces.

UNIT 5: MASS TRANSFER

Part-A (2 Marks with Answers)

1. What is meant by mass transfer? Give examples.

The process of transfer of mass as a result of the species concentration difference in a mixture is known as mass transfer.

Ex:

i. Humidification o air in a cooling tower

ii. Evaporation of petrol in the carburetor of an IC engine.

iii. The transfer of water vapor into dry air.

2. What are the modes of mass transfer?

a. Diffusion mass transfer

b. Convective mass transfer

3. Differentiate molecular diffusion and eddy diffusion.

Molecular Diffusion

The transport of water on a microscopic level as a result of diffusion from a region of higher concentration to the region of lower concentration in a mixture of liquids or gases is known as molecular diffusion.

Eddy Diffusion

When one of the diffusion fluids is in turbulent motion, eddy diffusion will takes place.

4. What is convective mass transfer?

It is the process of mass transfer that will occur between a surface and a fluid medium when they are at different concentrations.

5. State Flick's law of diffusion.

It states that molar flux of an element per unit area is directly proportional to concentration gradient.

$$M_a/A = -D_{ab} \, (dc_c/dx)$$

Where

M_a/A = Molar flux – Kg-mole/sm^2

D_{ab} = Diffusion coefficient of species a and b in m^2/s

dc_c/dx = Concentration gradient in kg/m^3

6. **What is free convective mass transfer?**

If the fluid motion is produced due to change in density resulting from concentration gradients, the mode of mass transfer is said to be free or natural convective mass transfer.

Ex: Evaporation of alcohol.

7. **What is forced convective mass transfer?**

If the fluid motion is artificially created by means of an external force like a blower or fan, that type of mass transfer is known as forced convective mass transfer.

Ex: The evaporation of water from an ocean when air blows over it.

8. **Define Schmidt number and Scherwood number.**

Schmidt Number

It is defined as the ratio of the molecular diffusivity of momentum to the molecular diffusivity of mass.

Scherwood Number

It is defined as the ratio of concentrations gradients at the boundary.

9. **Define mass concentration.**

Mass concentration or mass density is defined as the mass of a component per unit volume of the mixture. It is expressed in terms of kg/m^3

10. **Define molar concentration.**

Molar concentration or molar density is defined as the number of molecules of a component per unit volume of the mixture. It is expressed in terms of $kg\text{-}mole/m^3$

11. **Define mass fraction.**

The mass fraction is defined as the ratio of the mass concentration of species to the total mass density of the mixture.

12. **Define mole fraction.**

The mole fraction is defined as the ratio of mole concentration of a species to the total molar concentration.

QUESTIONS

UNIT 1

Conduction

Problems

1. A long aluminium cylinder 5 cm in diameter and initially at 200°C is suddenly exposed to a convection environment at 70°C and h = 525 W/m²K. Determine the temperature at a radius of 1.25 cm and the heat lost per unit length 1 minutes after the cylinder is exposed to environment.

   Take ρ = 2700 kg/m³, C = 0.9 kJ/kgK, k = 215 W/mK, α = 8.4 x 10⁵ m²/s.

2. A slab of rubber of thickness 40 cm, initially at a uniform temperature of 300°C. It is exposed to air at 30°C, the convection co-efficient being 240 kJ/hr m² °C. Assuming that it is a large slab, find the mid plane temperature after 15 minutes.

3. A steel plate (α=1.25 x 10⁻⁵ m²/s, ρ = 7833 kg/m³, Cp = 465 J/kg °C, k = 43 W/m° C) thickness 5 cm initially at uniform temperature of 200°C is suddenly immersed in an oil both at 20°C. The convection heat transfer co-efficient between the fluid and the surface is 500 W/m² °C. How long will it take for the center plane to cool to 100°C.

4. A steel pipe line (K = 50 W/mK) of inner diameter 100 mm and outer diameter 110 mm is to be covered with two layers of insulation each having a thickness of 50 mm. The thermal conductivity of the first insulation material is 0.06 W/mK and that of the second is 0.12 W/mK. Calculate the loss of heat per meter length of pipe and the interface temperature between the two layers of insulation when the temperature of the inside tube surface is 250°C and that of the outside surface of the insulation is 50°C.

5. A metallic sphere of radius 10 mm is initially at a uniform temperature of 400°C. It is heat treated by first cooling it in air (h = 10W/m²K) at 20°C until its central temperature reaches 335°C it is then quenched in a water bath at 20°C with h = 6000 W/m²K until the centre of the sphere cools from 335°C to 50°C. Compute the time required for cooling in air and water for the following physical properties of the sphere density = 3000 kg/m³, specific heat = 1000 J/kgK, thermal conductivity = 20 W/mK, thermal diffusivity = 6.66 x 10⁻⁶ m²/sec.

6. An aluminium fin (k = 200 W/mK) 3 mm thick and 7.5 cm long protrudes from a wall at 300°C. The ambient temperature is 50°C with h = 10W/m²K. Compute heat loss from the fin per unit depth of the material. Also calculate its efficiency and effectiveness.

7. A one meter long, 5 cm diameter cylinder placed in atmosphere at 40°C is provided with 12 longitudinal straight fins (k = 75.6 W/mK). The fins are 0.8 mm thick and protrude 2.5 cm from the cylinder surface. The heat transfer co-efficient is 23.25 W/m²K. Calculate the rate of heat transfer if the surface temperature is 150°C.

8. A steel tube with 5 cm ID, 7.6 cm OD and K = 15 W/m °C is covered with an insulated covering of thickness 2 cm and K = 0.2 W/m°C. A hot gas at 330°C with h = 400 W/m² °C flows inside the tube. The outer surface of the insulation is exposed to cooler air at 30°C with h = 60 W/m² °C. Calculate the heat loss from the tube to the air for 10 m of the tube and the temperature drops resulting from the thermal resistances of the hot gas flow, the steel tube, the insulation layer and the outside air.

9. An exterior wall of a house may be approximated by a 0.1 m layer of common brick. (K = 0.7 W/m°C) followed by a 0.04 m layer of gypsum plaster (K = 0.48 w/m°C). What thickness of loosely packed rock wool insulation (K = 0.065 W/m°C) should be added to reduce the heat loss or gain through the wall by 80%.

10. The temperature distribution across a large concrete slab (K = 1.2 W/m°C, α = 1.77 x 10^{-3} m²/h) 500 mm thick heated from one side as measured by thermocouples approximates to the relations $l = 60 - 5x + 12x^2 + 20x^3 - 15x^4$ where l is in °C and x is in meters.
    Considering an area of 5 m² compute:

    i.   The heat entering and leaving the slabs in unit time.

    ii. The heat energy stored in unit time.

    iii. The rate of temperature change at both sides of the slabs.

    iv. The point where the rate of heating or cooling is maximum.

UNIT 2

Convection

1. Air at 20° C at 3 m/s flows over a thin plate of 2 m long and 1 m wide. Estimate the boundary layer thickness at trailing edge, total drag force, mass flow of air between x = 30 cm and x = 80 cm. Take γ = 15 x 10^{-6} m²/s, ρ = 1.17 kg/m³.

2. Atmospheric air at 150°C flows with a velocity of 1.25 m/s over a 2 m long flat plate whose temperature is 25°C. Determine the average heat transfer coefficient and rate of heat transfer for a plate width of 0.5 m.

3. Caster oil at 25°C flows at a velocity of 0.1 m/s part a flat plate, in certain process.
    If plate is 4.5 m long and is maintained at uniform temperature of 95°C, calculate the following:

i. The hydrodynamic and thermal boundary layer thickness on one side of plate.

ii. The total drag force per unit width on one side of plate.

iii. Local heat transfer coefficient at the trailing edge and heat transfer rate. Properties of oil at 60°C: ρ = 956.8 kg/m^3, α = 7.2 x 10^{-8} m^2/s, K = 0.213 W/mK, γ = 0.65 x 10^{-4} m^2/s.

4. An aero-plane flies with speed of 450 km/h at a height where the surrounding air has temperature of 1°C and pressure of 65 cm of Hg. The aero-plane wing idealized as flat plate 6 m long, 1.2 m wide is maintained at 19°C. If flow is made parallel to 1.2 m width. Calculate (i) Heat loss from wing; (ii) Drag force on wing.

5. In straight tube of 50 mm diameter, water is flowing at 15 m/s. The tube surface temperature is maintained at 60°C and flowing water is heated from inlet temperature 15°C to an outlet temperature of 45°C. Calculate the heat transfer coefficient from the tube surface to water and length of tube.

6. A flat plate 1 m wide and 1.5 m long is to be maintained at 90°C in air with free stream temperature of 10°C. Determine the velocity with which air must flow over flat plate along 1.5 m side, so that rate of energy dissipation from the plate is 3.75 kW. Take Cp = 1.007 kJ/kg$^\circ$C, μ = 2.03 x 10^{-5} kg/m-s, pr = 0.7

7. A steam pipe 10 cm outside diameter runs horizontally in a room at 23°C. Take outside surface temperature of pipe as 165°C. Determine the heat loss per metre length of pipe.

8. A thin 80 cm long and 8 cm wide horizontal plate is maintained at a temperature of 130°C in large tank full of water at 70°C. Estimate rate of heat input into the plate necessary to maintain the temperature of 130°C.

9. A hot plate 1.2 m wide, 0.35 m high and at 115°C is exposed to ambient air at 25°C. Calculate (i) Maximum velocity at 180 mm from the leading edge of plate; (ii) Boundary layer thickness at 180 mm from leading edge; (iii) Local heat transfer coefficient at 180 mm from leading edge; (iv) Average heat transfer coefficient over surface of plate; and (vi) Heat loss from plate and rise in temperature of air passing through boundary.

10. Cylindrical cans of 150 mm length and 65 mm diameter to be cooled from an initial temperature of 20°C by placing them in a cooler containing air at a temperature of 1°C and a pressure of 1 bar. Determine cooling rates when cans are kept in (i) horizontal position; and (ii) vertical position.

UNIT 3

Phase Change Heat Transfer and Heat Exchanger

1. Water at atmospheric pressure is to be boiled in a polished copper pan. The diameter of the pan is 350 mm and is kept at 115°C. Calculate the power of the burner, rate of evaporation in kg/hr and the critical heat flux.

2. Explain nucleate boiling and solve the following:

   A wire of 1 mm diameter and 150 mm length is submeter horizontally in water at 7 bar. The wire carries a current of 131.5 ampere with an applied voltage of 2.16 volt. If the surface of the wire is maintained at 180°C. Calculate the heat flux and the boiling heat transfer coefficient.

3. The outer surface of a cylindrical vertical drum having 25 cm diameter is exposed to saturated steam at 1.7 bar for condensation. The surface temperature of the drum is maintained at 85°C. Calculate the following –

   a. Length of the drum

   b. Thickness of condensate layer to condense 65 kg / h of steam.

4. A condenser is to be designed to condense 600 kg/h of dry saturated steam at a pressure of 0.12 bar. A square array of 400 tubes, each of 8 mm diameter is to be used. The tube surface is maintained at 30°C. Calculate the heat transfer coefficient and the length of each tube.

5. A parallel flow heat exchanger has hot and cold water steam running through it, the flow rates are 10 and 25 kg/min respectively. Inlet temperatures are 75°C and 25°C on hot and cold sides. The exit temperature on the hot side should not exceed 50°C. Assume $h_i = h_o = 600$ W/m^{2}K. Calculate the area of heat exchanger using ε - NTU approach.

6. Hot exhaust gases enters a finned tube cross flow heat exchanger at 300°C and leave at 100°C, are used to heat pressurized water at a flow rate of 1 kg/sec from 35 to 125°C. The exhaust gas specific heat is approximately 1000 J/kgK and the overall heat transfer coefficient based on the gas side surface area is $U_h = 100$ W/m^{2}K. Determine the required gas side surface area A_h using the NTU method. Take C_{PC} at $T_C = 80$$^\circ$C is 4197 kJ/kgK and $C_{ph} = 1000$ J/kgK

7. In a double pipe counter flow heat exchanger, 10,000 kg/h of an oil having a specific heat of 2095 J/kgK is cooled from 80°C to 50°C by 8000 kg/h of water entering at 25°C. Determine the heat exchanger area for an overall heat transfer coefficient of 300 W/m^2.K. Take Cp for water as 4180 J/kgK.

8. Hot oil with a capacity rate of 2500 W/K flows through a double pipe heat exchanger. It enters at 360°C and leaves at 300°C. Cold fluid enters at 30°C and leaves at 200°C. If the overall heat transfer coefficient is 800 W/m²K, determine the heat exchanger area required for (a) Parallel flow and (b) Counter flow.

9. A counter flow concentric tube heat exchanger is used to cool engine oil (c = 2130 J/kg.K) from 160°C to 60°C with water, available at 25°C as the cooling medium. The flow rate of cooling water through the inner tube of 0.5 m diameter is 2 kg/s while the flow rate of oil through the outer annulus O.D. = 0.7 m is also 2 kg/s. If the value of the overall heat transfer coefficient is 250 W/m².K, how long must the heat exchanger be to meet its cooling requirement?

10. Saturated steam at 100°C is condensing on the shell side of a shell-and-tube heat exchanger. The cooling water enters the tubes at 30°C and leaves at 70°C. Calculate the effective log mean temperature difference if the arrangement is (i) counter flow, (ii) parallel flow and (iii) cross flow.

11. In a food processing plant water is to be cooled from 18°C to 6.5°C by using brine solution entering at an inlet temperature of -1.1°C and leaving at 2.9°C. What area is required when using a shell-and-tube heat exchanger with the water making one shell pass and the brine making two tube passes? Assume an average overall heat transfer coefficient of 850 W/m².K and a design heat load of 6000 W.

12. Water at the rate of 4 kg/s is heated from 38°C to 55°C in a shell-and-tube type heat exchanger. The water is to flow inside tubes of 2 cm diameter with an average velocity of 35 cm/s. Hot water available at 95°C and at the rate of 2.0 kg/s is used as the heating medium on the shell side. If the length of tubes must not be more than 2 m calculate the number of tube passes, the number of tubes per pass and the length of the tubes for one pass shell, assuming U_o = 1500 W/m²K

13. A refrigerator is designed to cool 250 kg/h of hot liquid of specific heat 3350 J/kgK at 120°C using a parallel flow arrangement. 1000 kg/h of cooling water is available for cooling purposes at a temperature of 10°C. If the overall heat transfer coefficient is 1160 W/m².K and the surface area of the heat exchanger is 0.25 m², calculate the outlet temperatures of the cooled liquid and water and also the effectiveness of the heat exchanger.

14. Water enters a counter flow, double pipe heat exchanger at 15°C, flowing at the rate of 1300 kg/h. It is heated by oil (Cp = 2000 J/kg.K) flowing at the rate of 550 kg/h from the inlet temperature of 94°C. For an area of 1m² and an overall heat transfer coefficient of

1075 W/m^{2}K determine the total heat transfer and the outlet temperatures of water and oil.

15. A 1-shell-2 tube pass steam condenser consists of 3000 brass tubes of 20 mm diameter. Cooling water enters the tubes at 20°C with a mean flow rate of 3000 kg/s. The heat transfer coefficient for condensation on the outer surfaces of the tubes is 15500 W/m^2.K. If the heat load of the condenser is 2.3 x 10^8 W when the steam condenses at 50°C determine (a) the outlet temperature of the cooling water; (b) the overall heat transfer coefficient; (c) the tube length per pass using the NTU method; (d) the rate of condensation of steam if h$_{fs}$ = 2380 kJ/kg.

16. Water enters a cross flow heat exchanger (both fluids unmixed) at 5°C and flows at the rate of 4600 kg/h to cool 4000 kg/h of air that is initially at 40°C. Assume the U value to be 150 W/m^2.K. For an exchanger surface area of 25 m^2, calculate the exit temperature of air and water.

UNIT 4

Radiation

1. Calculate following of an industrial furnace in form of black body and emitting radiation at 2500°C
    - (i)   Monochromatic emissive power at 1.2 μm
    - (ii)   Wavelength at emissive power is maximum
    - (iii)   Maximum emissive power
    - (iv)   Total emissive power.

2. What is black body? A 20 cm diameter spherical ball at 527°C is suspended in air. The ball closely approximates as a black body. Determine total black body emissive power, spectral body emissive power at wavelength 3 μm.

3. A pipe carrying steam having an outside diameter of 20 cm runs in a large room and is exposed to air at temperature of 30°C. The pipe surface temperature is 400°C. Calculate the loss of heat to the surroundings per metre length of pipe due to thermal radiation. The emissivity of the pipe surface is 0.8. What would be the loss of heat due to radiation if the pipe is enclosed in a 40 cm diameter brick conduct of emissivity 0.91.

4. Consider two large parallel plates one at T_1 = 1000°K with emissivity ε_1 = 0.8 and the other at T_2 = 500°K with emissivity ε_2 = 0.4. An aluminium radiation shield with a emissivity (both sides) ε_3 = 0.2 is placed between the plates. Calculate the percentage reduction in the heat transfer rate as a result of the radiation shield.

5. Two very large parallel planes with emissivity 0.3 and 0.8 exchange heat by radiation. Find the percentage reduction in heat transfer when a polished aluminium radiation shield of emissivity = 0.04 is placed between them.

6. Consider a cylindrical furnace with outer radius = 1 m and height = 1 m. Top surface and base have emissivity 0.8 and 0.4 and maintained at temperature of 700 K and 500 K. Side surface is approximately a black body and maintained at temperature 400 K. Find net radiation heat transfer at each surface.

7. Emissivity of two large parallel plates at 800°C and 300°C are 0.3 and 0.5. Find net radiant heat exchange per square meter. Find % reduction in heat transfer when polished aluminium radiation shield ε = 0.05 is placed between them.

8. Two parallel plates of size 1 m x 1 m are spaced 0.5 m apart are located in a very large room the walls of which are maintained at temperature of 27°C. One plate is maintained at a temperature of 900°C and other at 400°C. Their emissivity are 0.2 and 0.5. If plates exchange between themselves and surroundings. Find net heat transfer to each plate and to room. Consider only plate surfaces facing each other.

9. A chamber is filled with a gas mixture at a pressure of 2 atm and 1000°C. The gas mixture is transparent to radiation except CO_2 whose partial pressure is 0.3 atm. Assuming a mean beam length of 1.2 m, estimate the emissivity of the gas volume.

10. Determine the view factor F_{1-2} and F_{2-1} for the figure shown below:

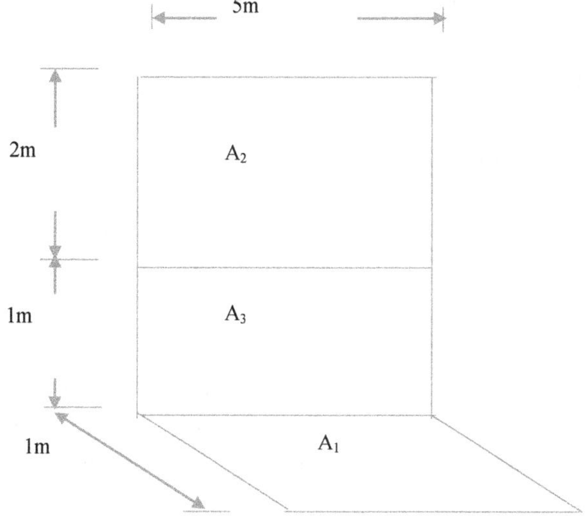

Unit 5

Mass Transfer

1. A mixture of O_2 and N_2 with their partial pressure in the ratio 0.21 to 0.79 is in a container at 25°C. Calculate the molar concentration, the mass density, and the mass fraction of each components for a total pressure of 1 bar. What would be the average molecular weight of the mixture?

2. The molecular weight of the two components A and B of a gas mixture are 24 and 48 respectively. The molecular weight of a gas mixture is found to be 30. If the mass concentration of the mixture is 1.2 kg/m³, determine the following:

   (i) Density of component A and B ii) Molar Fractions

   (iii) Mass fractions   iv) Total pressure if the temperature of the mixture is 290 K.

3. A thin plastic membrane separates hydrogen from air. The molar concentrations of hydrogen in the membrane at the inner and outer surfaces are determined to be 0.045 and 0.002 k-mol/m³, respectively. The binary diffusion coefficient of hydrogen in plastic at the operation temperature is 5.3×10^{-10} m²/s. Determine the mass flow rate of hydrogen by diffusion through the membrane under steady conditions if the thickness of the membrane is (1) 2 mm and (2) 0.5 mm.

4. Oxygen at 25°C and pressure of 2 bar is flowing through a rubber pipe of inside diameter 25 mm and wall thickness 2.5 mm. The diffusivity of O_2 in rubber is 3.12×10^{-3} kg-mole / m³-bar. Find the loss of O_2 by diffusion per meter length of pipe.

5. Two large tanks, maintained at the same temperature and pressure are connected by a circular 0.15 m diameter direct, which is 3 m in length. One tank contains a uniform mixture of 60 mole % ammonia and 40 mole % air, the other tank contains a uniform mixture of 20 mole % ammonia and 80% mole air. The system is at 273 K and 1.013×10^5 Pa. Determine the rate of ammonia transfer between the two tanks, assuming a steady state mass transfer.

6. A 3 cm diameter Stefan tube is used to measure the binary diffusion coefficient of water vapor in air at 20°C at an elevation of 1600 m where the atmospheric pressure is 83.5 kPa. The tube is partially filled with water, and the distance from the water surface to the open end of the tube is 40 cm. Dry air is blown over the open end of the tube so that water vapour rising to the top is removed immediately and the concentration of vapour at the

top of the tube is zero. In 15 days of continuous operation at constant pressure and temperature, the amount of water that has evaporated is measured to be 1.23 g. Determine the diffusion coefficient of water vapour in air at 20°C and 83.5 kPa.

7. An open pan of 150 mm diameter and 75 mm deep contains water at 25°C and is exposed to atmosphere air at 25°C and 50% RH. Calculate the evaporation rate of water in grams per hour.

8. Dry air at 15°C and 92 kPa flows over a 2 m long wet surface with a free stream velocity of 4 m/s. Determine the average mass transfer coefficient.

9. Air at 25°C and atmospheric pressure flows in a 10 mm diameter tube of 1 m length with a velocity of 3 m/sec. The inside surface of the tube contains a deposit of naphthalene. Determine the average mass transfer coefficient.

   Take diffusion coefficient $D_{ab} = 0.6 \times 10^{-5}$ m²/sec.

10. Air at 35°C and 1 atm flows over a wet flat plate 50 cm long with a velocity of 30 m/s. Calculate the mass transfer coefficient of water vapour in air at the end of the plate.